Medical Intelligence Unit

Skin Substitute Production by Tissue Engineering: Clinical and Fundamental Applications

Mahmoud Rouabhia, Ph.D.
Department of Surgery,
Faculty of Medicine
Department of Biochemistry,
Faculty of Sciences and Engineering
Laval University
Laboratoire des grands brûlés/LOEX
Québec, QC, Canada

CHAPMAN & HALL
I(T)P An International Thomson Publishing Company
New York • Albany • Bonn • Boston • Cincinnati • Detroit • London • Madrid • Melbourne • Mexico City • Pacific Grove • Paris • San Francisco • Singapore • Tokyo • Toronto • Washington

LANDES BIOSCIENCE
Austin, Texas
U.S.A.

Medical Intelligence Unit

SKIN SUBSTITUTE PRODUCTION BY TISSUE ENGINEERING:
CLINICAL AND FUNDAMENTAL APPLICATIONS

R.G. LANDES COMPANY
Austin, Texas, U.S.A.

Printed in the U.S.A.

Please address all inquiries to the Publishers:
R.G. Landes Company, 810 South Church Street, Georgetown, Texas, U.S.A. 78626
Phone: 512/ 863 7762; FAX: 512/ 863 0081

North American distributor:

Chapman & Hall, 115 Fifth Avenue, New York, New York, U.S.A. 10003

CHAPMAN & HALL

U.S. and Canada ISBN: 0-412-13971-5

While the authors, editors and publisher believe that drug selection and dosage and the specifications and usage of equipment and devices, as set forth in this book, are in accord with current recommendations and practice at the time of publication, they make no warranty, expressed or implied, with respect to material described in this book. In view of the ongoing research, equipment development, changes in governmental regulations and the rapid accumulation of information relating to the biomedical sciences, the reader is urged to carefully review and evaluate the information provided herein.

Library of Congress Cataloging-in-Publication Data

Skin substitute production by tissue engineering / editor, Mahmoud Rouabhia.
p. cm. — (Medical intelligence unit)
Includes bibliographical references and index.
ISBN 1-57059-447-3 (alk. paper)
1. Skin--Cultures and culture media. 2. Skin--transplantation. 3. Keratinocytes.
4. Animal cell biotechnology. I. Rouabhia, Mahmoud, 1957- . II. Series.
[DNLM: 1. Skin transplantation. 2. Skin--surgery. 3. Skin, Artificial.
4. Biomedical Engineering. 5. Tissue culture. WO 610 S6278 1997]
QP88.5.S576 1997
617.4'770592—dc21
DNLM/DLC 97-6738
for Library of Congress CIP

Publisher's Note

R.G. Landes Bioscience Publishers produces books in six Intelligence Unit series: *Medical, Molecular Biology, Neuroscience, Tissue Engineering, Biotechnology* and *Environmental.* The authors of our books are acknowledged leaders in their fields. Topics are unique; almost without exception, no similar books exist on these topics.

Our goal is to publish books in important and rapidly changing areas of bioscience for sophisticated researchers and clinicians. To achieve this goal, we have accelerated our publishing program to conform to the fast pace at which information grows in bioscience. Most of our books are published within 90 to 120 days of receipt of the manuscript. We would like to thank our readers for their continuing interest and welcome any comments or suggestions they may have for future books.

Shyamali Ghosh
Publications Director
R.G. Landes Company

Dedication

This book is dedicated to my parents in Tunisia; to my spouse, Jamila and my daughter, Dounia whose support have made this edition possible and exciting.

This book is also dedicated to all the members of the Laboratoire de Recherche des Grands Brûlés/LOEX team at The Saint-Sacrement Hospital, Laval University who, over the years, have provided valuable help and a superb and enriching ambience in which my research projects were undertaken and progressed continuously.

CONTENTS

Section I
Clinical Use of Engineered Skin Substitutes

Section II
Experimental Use of Engineered Skin Substitutes

ABBREVIATIONS

Chapter 1

APC	antigen presenting cell
CD	cluster of differentiation
CSF	colony stimulating factor
DDC	dendritic dermal cell
ECF-A	eosinophil chemotactic factor of anaphylaxis
GM-CSF	granulocyte macrophage-CSF
HETE	hydroxyeicosatetraenoic acid
HLA	human leukocyte antigen
IL	interleukin
LC	Langerhans cell
LTB	leukotriene-B
LTC	leukotriene-C
MHC	major histocompatibility complex
NCF	neutrophil chemotactic factor
PAF	platelet activating factor
PGD	prostaglandin-D
TCR	T cell receptor
Th	T helper

Chapter 3

BPE	bovine pituitary extract
CAE	cultured autologous epidermis
DMEM	dubelco modified eagle's medium
EGF	epithelial growth factor
PBS	phosphate-buffered saline
TBSA	total body surface area

Chapter 5

FITC	fluorescein-isothiocyanate
MHC	major histocompatibility complex
OCT	optimal cutting temperature
PBS	phosphate buffer saline
PE	phycoerytrin

Chapter 7

APC	antigen presenting cell
DS	dermal substitute
EGF	epidermal growth factor
IL	interleukin
LC	Langerhans cell
SS	skin substitute
Th1	T helper-1
Th2	T helper-2
TNF-α	tumor necrosis factor-α
UV	ultraviolet
UVR	UV radiation

Chapter 8

FITC	fluorescein-isothiocyanate
FL1	fluorescence intensity
FSC	forward light scatter
K	keratin
kDa	kilo Dalton
PBS	phosphate buffered saline

EDITOR

Mahmoud Rouabhia, Ph.D.
Department of Surgery,
Faculty of Medicine
Department of Biochemistry,
Faculty of Sciences and Engineering
Laval University
Laboratoire des grands brûlés/LOEX
Québec, QC, Canada
Chapters 1, 2, 5, 6, 7, 10

CONTRIBUTORS

François A. Auger, M.D., Director
Laboratoire des
grands brûlés/LOEX
Department of Surgery
Laval University
Québec, QC, Canada
Chapter 9

François Berthod, Ph.D.
Department of Surgery
Laval University
Laboratoire des
grands brûlés/LOEX
Québec, QC, Canada
Chapters 2, 9

Steven T. Boyce, Ph.D.
Department of Surgery
University of Cincinnati
Medical Center
Cincinnati, Ohio, U.S.A.
Chapter 4

Fabienne Braye, M.D.
Department of Surgery
Centre des brûlés
Hôpital Edouard Herriot
Lyon, France
Chapter 3

Nadine Chouinard, B.Sc.
Department of Biochemistry
Faculty of Sciences
and Engineering
Laval University
Laboratoire des
grands brûlés/LOEX
Québec, QC, Canada
Chapter 7

Odile Damour, Ph.D.
Laboratoire des substituts cutanés
Hôpital Edouard Herriot
Lyon, France
Chapter 3

Anne Fabreguette, Pharm.D.
Laboratoire des substituts cutanés
Hôpital Edouard Herriot
Lyon, France
Chapter 3

Jean-Louis Foyatier, M.D.
Department of Surgery
Hôpital Saint-Luc
Lyon, France
Chapter 3

Julie Fradette, M.Sc.
Department of Surgery
Laval University
Laboratoire des
grands brûlés/LOEX
Québec, QC, Canada
Chapter 8

Lucie Germain, Ph.D
Department of Surgery
Laval University
Laboratoire des
grands brûlés/LOEX
Québec, QC, Canada
Chapter 8

Marie-Josée Godbout
Faculty of Medicine
Laval University
Laboratoire des
grands brûlés/LOEX
Québec, QC, Canada
Chapter 8

Roger Leblanc, Ph.D.
Department of Chemistry
University of Miami
Coral Gables, Florida, U.S.A.
Chapter 7

Hui Li, Ph.D.
Department of Surgery
Laval University
Laboratoire des
grands brûlés/LOEX
Québec, QC, Canada
Chapter 8

Martine Michel, Ph.D.
Research Center in Oral Biology
Health Sciences Center
University of Washington
Seattle, Washington, U.S.A.
Chapter 8

Paul Petit, Pr. M.D.
Laboratoire des substituts cutanés
Hôpital Edouard Herriot
Lyon, France
Chapter 3

Patricia Rousselle, Ph.D.
Institut de biologie et de chimie
des protéines
CNRS UPR 412
Lyon, France
Chapter 3

Séverine Vissac, Pharm.D.
Laboratoire des substituts cutanés
Hôpital Edouard Herriot
Lyon, France
Chapter 3

Wen Xu, Ph.D.
Department of Surgery
Laval University
Laboratoire des
grands brûlés/LOEX
Québec, QC, Canada
Chapter 8

PREFACE

Usually, the skin does not enjoy important consideration by its occupant. Humans always give this organ valuable consideration only after partial or complete damage. Indeed, skin functions are multiple and vital. Thus, neither humans nor animals are able to survive without this organ.

As emphasized by the statement "Beauty is only skin-deep", the skin is the human mirror in our society today, and the body image is a basic part of personal identity. Skin is the organ most exposed to harmful environmental effects and the first biological organ to be extensively damaged by irradiation, cold or heat.

Major burn injuries evoke strong emotional responses in almost all persons and health professionals who are confronted by the specter of pain, deformity and potential death associated with significant skin loss. Thus, from family members to physicians, the challenge to help burn patients survive and, then, accept their new personal identity and social mirror is constantly present.

This book outlines in the first section different methods and materials used to clinically achieve a safe temporary closure of the open wound when sufficient autologous skin is not available. The most common biological dressing (allogeneic and xenogeneic skin), synthetic and biosynthetic materials used for burn treatment and coverage are also discussed. Chapter 3 and 4 of this section outline some interesting clinical results obtained over the last decade in different burn units where damaged skin is replaced permanently using in vitro engineered epidermis. In the second section of this book, various experimental studies are documented using in vitro engineered skin. Indeed, as burn treatment based on engineered epidermis requires several weeks for the patient's keratinocyte growth before damaged skin replacement, I devised a new tissue engineering method (chimeric cultures) to overcome this drawback. Using engineered chimeric epidermis, encouraging results are presented in chapters 5 and 6 of this book. In other chapters, we also point out the potential use of this multidisciplinary and interdisciplinary field (tissue engineering) for a better understanding of cutaneous stem cell specificity. We tried to shed light on some of the mechanisms involved in the deleterious effects of ultraviolet light on cutaneous cells and extracellular matrix. Finally, in vitro engineered skin substitutes present promising tools for valuable pharmacological and toxicological studies. Our hope is that the current and future research in the area of tissue engineering will lead to improved care of all burned patients and open the horizon for engineering other organs.

This book was made possible with the support of many individuals. I would like to thank Prof. S. William, A. Gunn and Dr. Mark R. Pittelkow for their critical reading of some chapters published in this book. I would like to also thank the clinicians, Drs. Félix-André Têtu, Alphonse Roy, Normand Houle and Georges Hébert, for providing tissue specimens. I wish to express sincere appreciation to PharMingen Canada and VWR Canlab for their financial contributions to this book. I want to thank the Medical Research Council of Canada, the "Fonds de Recherche en Santé du Québec", the "Fondation des Pompiers de Québec pour les Grands Brûlés", and the "Fondation de l'Hôpital du Saint-Sacrement" for their critical financial support of the clinical and basic science research efforts in our LRGB/LOEX over the years. Finally, special thanks go to the audio-visual service at The Saint-Sacrement Hospital, mainly Mr. Claude Marin for his understanding, availability and help.

Mahmoud Rouabhia, Ph.D.
Surgery Department,
Laval University
February, 1997

SECTION I

Clinical Use of Engineered Skin Substitutes

CHAPTER 1

Structural and Functional Complexity of the Skin

Mahmoud Rouabhia

INTRODUCTION

The skin is an organ of sharp contrasts. It is considered the largest and most visible organ comprising a sheet-like investment of the whole body that adapts admirably to its contours and neatly conforms to the movements of the organism within. The average adult human skin exceeds $2m^2$ in area, with millions of hairs and glands. Its uniform appearance belies its great variation from one area of the body to another.[1,2]

Skin is the appropriate interface between the organism and its environment. It is adapted to withstand several physical, chemical, microbial etc., variables.

Anatomically, the skin is divided into three major tissue layers: the epidermis, dermis and hypodermis. All of these cutaneous structures intercommunicate giving a physiologically functional cutaneous tissue.

SKIN STRUCTURE

THE EPIDERMIS

The epidermis is the outermost part of the skin. This cutaneous structure is composed mainly of cells, with very little extracellular matrix.[3,4] It is a continually renewing structure composed of several layers (Fig. 1.1).

Skin Substitute Production by Tissue Engineering: Clinical and Fundamental Applications, edited by Mahmoud Rouabhia.

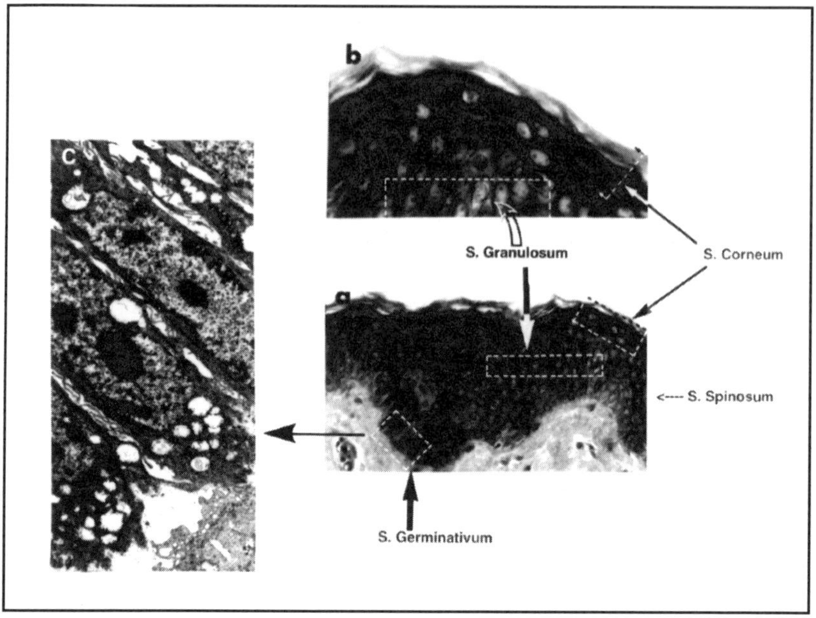

Fig. 1.1. Histological organization of epidermal layers. (a) Light microscopy of full-thickness epidermis with the upper part of the dermis. X200 in the original magnification. (b) Light microscopy of upper layers of the skin (stratum corneum and granulosum). X300 in the original magnification. (c) Transmission electron microscopy of the stratum germinativum. X4000 in the original magnification.

The Stratum Corneum

The stratum corneum is the outer layer of the epidermis. It is comprised of large, flattened and polyhedral-shaped cells (Fig. 1.1a,b). Those on the surface have more exaggerated flattening than the deeper ones. Cells are linked by desmosomes, overlapping at their lateral margins, and interlock with cells of deep or superficial layers by ridges and grooves of the surface.[5,6] Keratin represents as much as 80% of the cornified cell content.[7] In this layer there is also filaggrin.[8] Remnants of organelles, especially profiles of membranes and melanin pigment, are also occasionally retained within the cells. Few enzymes have been identified in cornified cells.[9] These enzymes could be involved in remodeling intercellular lipids and carbohydrate cell surface components. There is a subtle structural organization of the stratum corneum. Indeed, when hydrated or swollen, cornified cells appear to align in columns. It is proposed that the order of these columns is generated by a specific and intrinsic organization of cells in the viable layers and that this nonrandom order is important to the functional properties of the epidermis. The stratum corneum layer lies on the stratum granulosum.

The Stratum Granulosum

This structure contains two to three layers of nucleated cells (Fig. 1.1a,b). These flattened cells contain distinctive cytoplasmic inclusions and keratohyalin granules.[10,11] Two other major proteins are found in the stratum granulosum: filaggrin and involucrin.[12,13] Keratohyalin granules are linked with epidermal cell differentiation. The absence of keratohyalin granules and/or their structural abnormalities are features of several keratinization disorders. In the stratum granulosum, granular cells not only synthesize and cross-link new proteins such as filaggrin and involucrin to keratinization, but they also play roles in programmed cell death (apoptosis). During this phase, granular cells lose their nucleus and all contents except the keratin filaments and electron-dense matrix material.[14] At this stage, granular cells are, in reality, cornified cells. Beneath the stratum granulosum, we histologically distinguish the stratum spinosum.

The Stratum Spinosum

Cells in this epidermal region (Fig. 1.1a) present a cytologically polyhedral configuration and become progressively flattened as they move toward the granulosum and corneum strata, respectively. Cells of all spinous layers contain visible keratin filaments organized into bundles around the nucleus. These cells also contain typical cellular organelles, microfilaments and melanosome complex.[7] However, there are differences between upper and lower spinous layers. Upper layers correlate with progressive stages of cell keratinization and lower layers are composed of suprabasal cells which have a polyhedral shape and rounded nucleus. Stratum spinosum seems to contain active keratinocytes. Indeed, there is strong evidence of an endocytosis function performed by keratinocytes in the stratum spinosum. This function facilitates the elimination of nondegradable materials present in the dermis.[15,16] Upper spinous and granular cell layers contain small, oval-shaped lamellar granules. These granules carry neutral sugars conjugated to protein and lipid and several hydrolases. They enhance the barrier properties of the stratum corneum.[17]

The Stratum Germinativum

The last element of the epidermis is the stratum germinativum (basal cell layer). It is the deepest and only epidermal structure that directly communicates with the dermis via the basement membrane zone. The stratum germinativum contains high mitotically active keratinocytes (Fig. 1.1a,c). These cells have an oval nucleus, prominent nucleolus, little heterochromatin and typical "housekeeping" cellular organelles with enzyme systems. Basal cells contain keratin filaments organized into fine bundles around the nucleus, hemidesmosomes and desmosomes.[18] Other cytoskeletal elements found in basal cells are microfilaments and microtubules. Microfilaments participate in the upward movement of epidermal cells and replenish keratinocytes lost at the surface during normal epidermal turnover.[19,20] Once a basal keratinocyte leaves the basal layer, it needs at least 14 days to reach the stratum corneum, then transits through cornified layers and desquamation requires another 14 days. Basal cells are morphologically uniform but functionally heterogeneous with regard to their proliferative potential.[20] The epidermis maintains basal stem cells which are noncycling or divide very slowly. When divided, stem cells produce one daughter cell that will enter the cycling pool, undergo augmentation division and then differentiate. The other daughter cell remains in the basal layer as a stem cell. Stem cells can be stimulated to increase the number of cycling cells by wounding, carcinogens, diseases such as epidermal hyperplasia or alteration of normal proliferative mechanisms.[20]

As it is a nonvascularized structure, the epidermis maintains the proper proliferating and differentiating mechanisms through a tight interrelationship with the dermis via the basement membrane zone.[21] The basement membrane is an important interface which separates the epidermis from the dermis both physically and functionally.

THE DERMIS

During the fetal period, the dermis consists largely of loosely arranged mesenchymal cells embedded in a soft ground substance. In postnatal life, the dermis is an integrated system of fibrous, filamentous and amorphous connective tissue that facilitates vascular, nerve and cutaneous immune system networks. The dermal matrix contains fibrous and nonfibrous connective tissues. Fibrous molecules impart bulk, density, compliance, elasticity and tensile strength to the skin. Nonfibrous molecules permit cellular migration and physiologically regulate the osmotic properties of skin.[22,23]

The dermis contains two main anatomical portions, the papillary and the reticular regions. Papillary dermis is intimately associated with the epidermis via the basement membrane zone. Papillary dermis contains different protein forms (collagen and elastic fibers, enzymes, etc.) with a high proportion of type III collagen.[24,25] The reticular dermis represents the bulk of the dermis. It includes large bundles of collagen fibrils and mature elastic fibers.[24,25] The same collagens present in papillary dermis are also characteristic of reticular dermis. However, reticular dermis is composed primarily of type I collagen. The boundary between the papillary and reticular dermis is defined by a horizontal subpapillary plexus of arterioles and postcapillary venules.[22] Dermal cells are more abundant in the papillary than in the reticular dermis, although fibroblasts, macrophages and mast cells are found in the interstitium between collagen bundles.[22]

Dermis is a highly vascularized cutaneous structure with a vascular network consisting of distributing and collecting channels and microcirculatory plexuses of the papillary dermis and epidermal appendages. Also involved here is an intricate network of blood vessels that regulates body temperature and metabolic needs of the whole skin.[26,27]

Skin also contains nerve networks with somatic sensory and sympathetic autonomic fibers.[28] Lymphatic channels in the skin are important for the clearance of fluid, macromolecules, and cells from the dermis.[29] In normal skin, lymph channels are rarely present in the superficial papillary dermis. Finally, the skin contains a variety of appendages that produce different products: sebum, hair, and ecrine and apocrine sweat. Both dermis and epidermis are mixtures of different cell types.

CUTANEOUS CELLS

Epidermal Cells

The epidermis of an adult is a mixture of at least five cell types, three of which are of major importance: keratinocytes, melanocytes and Langerhans cells. Two additional types of cells (dendritic cells and Merkel cells) are occasionally found in the epidermis and oral mucosa.

Keratinocytes

Keratinocytes are the principal cells of the epidermis. They are of ectodermal origin and have the specialized function of producing keratin.[30] The epidermal organization into the above described strata reflects the phenotypic and somehow functional alteration of keratinocytes as they differentiate from proliferative to cornified cells. It was suggested that the keratinocyte goes through two phases: a synthetic and then a degradative phase on its way to becoming a horn cell. During the synthetic phase, the keratinocyte accumulates different fibrous proteins such as tonofilaments and keratins within its cytoplasm. Keratinocytes act as a mechanical protective barrier to the whole body[31] and play a major role in the immune functioning of the skin (its immune implication will be developed later on).

Melanocytes

Derived from the neural crest, melanocytes are the pigment-producing cells of the epidermis. They migrate into the fetal epidermis during the second month of embryonic development. In normal adult epidermis, melanocytes reside in the basal layer. Melanocytes maintain a relatively uniform distribution to a specific number of keratinocytes.[32] This is accomplished even though both are dividing cells. Melanocyte frequency is approximately 1 for every 10 basal keratinocytes. However, the number of melanocytes in the epidermis is the same regardless of race or color. What determines racial differences in skin color is the density and size of the melanosomes that are continuously synthesized by these melanocytes. Indeed, melanocytes of dark skin synthesize larger melanosomes compared to the smaller ones found in light skin. The size of the melanosome is the principal factor in determining how these proteins will be distributed within the keratinocytes. Larger melanosomes are individually dispersed within the cytoplasm of keratinocyes, while small melanosomes of light skin are packaged in a membrane-bound complex within the keratinocyte.[33] Melanosome phagocytosis by keratinocytes provides uniform distribution of the pigment to absorb light across the basal layer of the epidermis. It has been shown that sun exposure stimulates the melanocytes to produce larger melanosomes, thereby making the distribution of these proteins within keratinocytes resemble the pattern obtained in dark-skinned individuals.

Langerhans Cells

Langerhans cells (LC) originate from a mobile pool of bone marrow-derived precursor cells. They constitute only a minor subpopulation (1-5%) of epidermal cells. They preferentially populate stratified squamous epithelia of mammals, but are also found in mesenchymal tissues, such as the dermis, lymph nodes and thymus.[34] Phenotypically, LC display different markers and receptors (Table 1.1) and share several features with monocytes/macrophages and other dendritic cells present in lymphoid, as well as in nonlymphoid tissues.[35] Functionally, LC is involved in a large panoply of actions (Table 1.2), one of which is that LC are extremely potent stimulators of antigen-specific T cell activation. This physiological function initiates protective immune responses against endogenous and exogenous antigens.[36] LC critically participate in immune surveillance via different cytokine receptors (Table 1.3). Thus, quantitative

Table 1.1. Immunophenotypic features of cutaneous epidermal Langerhans cells

Human	Murine
CD45	CD45
MHC-I	MHC-I
MHC-II	MHC-II
Fc-IgG receptor-II	Fc-IgG receptor-II
CD11b/CD18	C3bi receptor
CD11c/CD18	T1a
CD54	F4/80
S-100	NLDC-145
Vimentin	J11d
HLA-D-Ii	Pgp-1
Lag	

Table 1.2. In vivo functions of Langerhans cells

Functions
Phagocytosis
Antigen presenting cell
Endogenous antigens
Exogenous antigens
Antigen transport (skin → lymph node)
Initiate antigen specific T cell responses
Initiate T cell-dependent B cell responses
Induction of secondary T cell responses
Participate in cutaneous immune surveillance
Antigen-specific immune tolerance
Involved in skin allograft rejection

Table 1.3. Cytokine receptors on epidermal Langerhans cells

Receptors
GM-CSF
M-CSF
IL-2 α-chain
IL-2 γ-chain
Type I IL-1

These receptors are found only on cultured LC

and/or qualitative impairment of these cells have deleterious consequences for the immunological defense of the host.[37]

Merkel Cells

Merkel cells are cutaneous cells with a specific ultrastructure distinguishing them from all other epidermal and dermal cells. They are characterized by several features. They present small cytoplasmic dense-core granules containing different neuropeptides, spiny or microvillus-like projections, intermediate filaments, desmosomes, lobular nucleus and generally synaptic contact with an axon terminal.[38]

Merkel cell distribution is variable and their density is body site dependent.[38,39] There are two prevailing hypotheses regarding the origin of Merkel cells: the neural crest and cutaneous origin hypotheses. However, the most plausible is the epidermal origin hypothesis. Functionally, Merkel cells derived from the neural crest are involved in slowly-adapting type I mechanoreception. However, it is still unclear whether or not these cells themselves act as mechanical transducers. Epidermal Merkel cells seem to stimulate local proliferation and differentiation of keratinocytes. Furthermore, epidermal Merkel cells may serve in maintaining normal epidermal structure by controlling the apoptotic deletion of excessively produced epidermal cells in the toad skin. Finally, the participation of Merkel cells in the hyperplastic, dysplastic and neoplastic changes of epidermal structures has been demonstrated.[38]

Dermal Cells

The constituents of the dermis are mesodermal except for nerves, which derive from the neural crest. Fetal (5 to 6 weeks of age) dermis is merely a pool of scattered, dendritic-shaped cells, which are the precursors of fibroblasts. Three months later, fibroblasts are mature and active cells. The dermis contains regular resident cells such as fibroblasts, macrophages and mast cells, and a circulating and/or recruited cell population that rallies T lymphocytes and dermal dendritic cells.

Fibroblasts

Dermis is the intrinsic component originating from cells of the superficial dermatome segments of somite mesoderm and from other mesenchymal areas of the body subjacent to the skin. The major cells differentiating from these anatomic sites are fibroblasts.[40,41] These cells are regular residents of the dermis. They are found in greatest density in the papillary region and around vessels. They are also present in the reticular dermis located in the interstices between collagen fiber bundles. Fibroblast morphology often suggests active synthetic activity; there are different profiles of dilated, rough endoplasmic reticulum and several Golgi complexes. Sites of intercommunication among the internal cytoskeleton, integral proteins of the plasma membrane and the extracellular matrix proteins were identified as focal densities along the plasma membrane of these fibroblasts.

Fibroblast attachment to the fibrous matrix is facilitated by fibronectin. Collagen and fibronectin have complementary binding sites as part of their molecular structures. Fibroblasts migrate among and on the surface of fiber bundles in the tissue but rarely do they invade or become trapped within them. They synthesize and secrete qualitatively and quantitatively different proteins that provide mechanical support to the skin (Fig. 1.2). Fibroblasts also synthesize and secrete different enzymes such as collagenase and gelatinase which participate in extracellular matrix regulation.[42]

Macrophages

Macrophages develop from bone-marrow precursor cells, which differentiate into promonocytes and monocytes in the blood and then into macrophages in the tissue. They are large (10-20 μm), actively mobile, phagocytic cells. Macrophages are distinguished from neutrophils by their unlobed nucleus and absence of specific granules characterizing these neutrophils. Macrophages are localized in different tissues such as lymph node, spleen, skin, liver (Kupffer cells), lung, and wherever foreign antigens invade the body. In the skin, the number of these macrophages increases after local inflammatory stress. While some macrophages are mobile, others are resident (Kupffer cells of the liver). Macrophages carry high amounts of major histocompatibility complex (MHC) class-II antigens and bear Fc and C3b receptors. They are active antigen-presenting cells (APCs). They engulf foreign particles and cellular debris into phagocytic vacuoles. These cells serve as part of the effector arm of humoral immunity and play active roles in cell-mediated immune responses.[43] Macrophages are

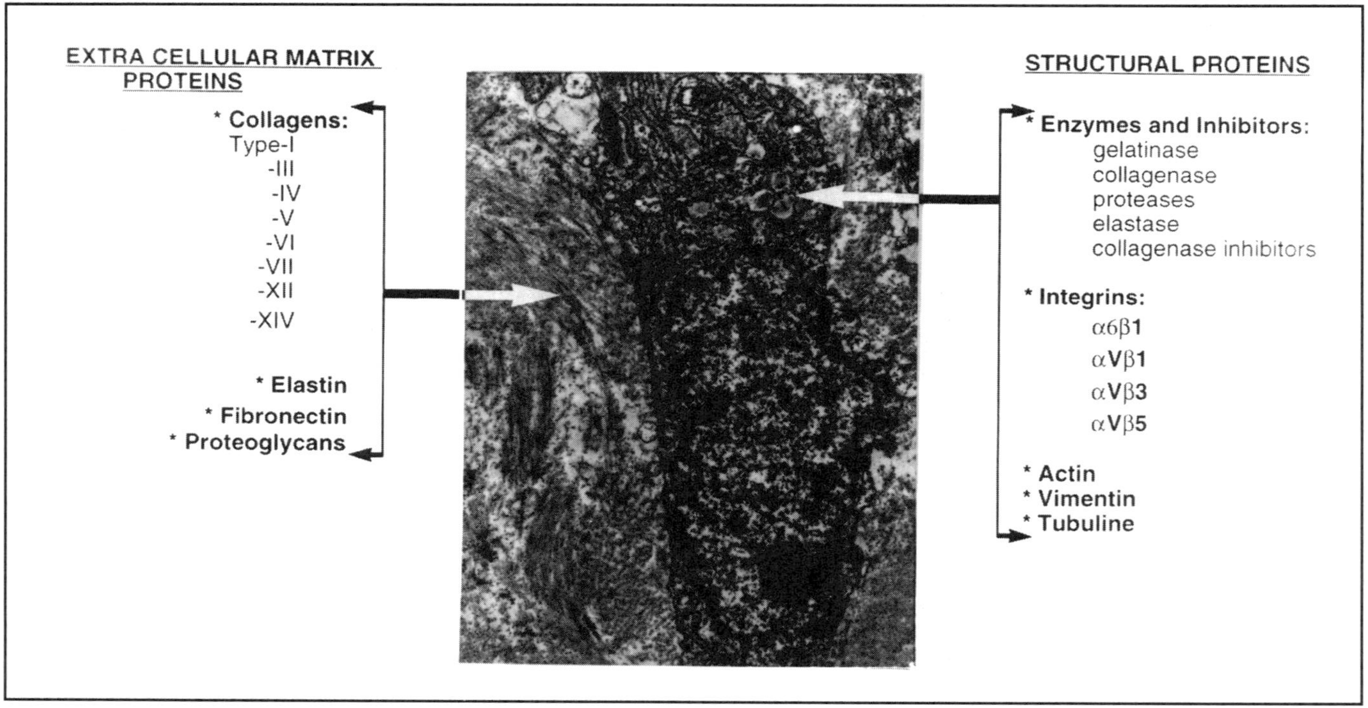

Fig. 1.2. Transmission electron microscopy of dermal fibroblasts. Note the abundant extracellular matrix fibers (collagen, elastic, etc.) surrounding the fibroblast. In the other part of this figure the different structural proteins were listed. X8000 in the original magnification.

involved in synthesis and secretion of hydrolytic enzymes, complement components and soluble factors modulating the immune response.

Mast Cells

Mast cells are bone-marrow derived cells. They are present in different tissues. These cells are present in all regions of the dermis, but are most commonly found in the upper dermis around vessels and epidermal appendages and in the subcutaneous fat.[44] By their anatomical location and ability to generate a variety of mediators (Table 1.4), tissue mast cells appear to play an important role in both physiological and pathological events.[44] Although their common site of origin appears to be bone marrow, in vivo and in vitro studies of mast cells indicate that the tissues in which these cells reside greatly influence their morphology, biochemical composition, expression of surface receptors and response to potential stimuli. Human mast cells from varied anatomical sites appear quite similar.[45] However, in vitro studies indicate that human mast cells vary in terms of their mediator content. Human skin tissue mast cells store significant amounts of active proteins. Additionally, in vitro studies suggest that human cutaneous mast cells are responsive to a variety of immunologic and nonimmunologic stimuli. Dermal mast cells are surrounded by fibronectin, which plays a major role in these cell anchors to the extracellular matrix at inflammatory sites where they proliferate and release different mediators.[45]

Table 1.4. Mediators released by mast cells located in normal human skin

Vasoactive Mediators	Chemotactic Mediators	Enzymes	After Stimulation
Histamine	ECF-A	Tryptase	IgE
PGD2	NCF	Chymase	C3a, C5a
LTB4	HHT	Arylsulfatase	Ionophore A23187
LTC4	HETE	Hexosaminase	Morphine
PAF		Glucuronidase	Substance P
		Charboxypeptidase	Gastrin

Dendritic Cells

In the past 10 years, several studies have demonstrated a dermal dendritic cell population which is different from the LC population. Dermal dendritic cells (DDC) are located in perivascular areas.[46] Morphologically, these cells exhibit a folded nucleus and highly ruffled, irregular surface. Their cytoplasm contains organelles involved in active cellular metabolism. As opposed to LC, DDC is devoid of Birbeck granules.[46] Phenotypically, these cells exhibit different markers (Fig. 1.3). The culture of cutaneous cells isolated from human skin confirmed the presence of these dendritic cells and lights up two other phenotypic markers which were the expression of large amounts of MHC class-II antigens and a platelet marker.[47] Using anti-CD36 monoclonal antibody, we recently, showed that these DDC could be localized in the epidermis mainly in the basal layer. They were nonadherent cells, representing about 2% and expressing cytokeratin-AE3.[47,48] At a functional level, DDC are potent stimulators of allogeneic mixed leukocyte reaction and powerful accessory cells in the oxidative mitogenesis assay. However, despite the interesting information in the literature about these cells, their in vivo functional role in the skin's immune system remains to be investigated.

T Cells

With the regular resident cells, the skin contains recruited and/or passenger cells such as lymphocytes. Mammalian skin lymphocytes are of T cell phenotype as defined by the surface expression of CD3 associated T cell receptor (TCR). Their numbers were variable from

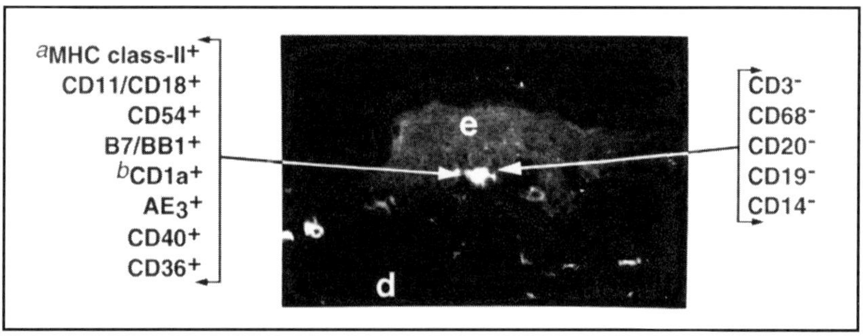

Fig. 1.3. Dendritic cells in normal human skin. Normal skin biopsies were stained with an anti-CD36 and goat anti-rat IgG-FITC monoclonal antibodies, then observed under fluorescence microscopy. Note the localization of CD36[+] cells and the presence of dendrites on them. X200 in the original magnification.

one study to another. T cells in human skin reside predominantly in the dermis and express TCR α/β rather then TCR γ/δ. In the dermis, T cells are located around postcapillary venules and the appendages (Fig. 1.4). Phenotypically, dermal T cells are $CD4^+$ and $CD8^+$ cells. However, intraepidermal T cells exhibit different phenotypes (Table 1.5) and are located within the basal layer. The vast majority of human

Table 1.5. Cutaneous T cell phenotypes

Marker	Dermis	Epidermis
CD2		+
CD3		+
CD4	+	–
CD8	+	+
TCR α/β	+	+
CD45RO	+	+

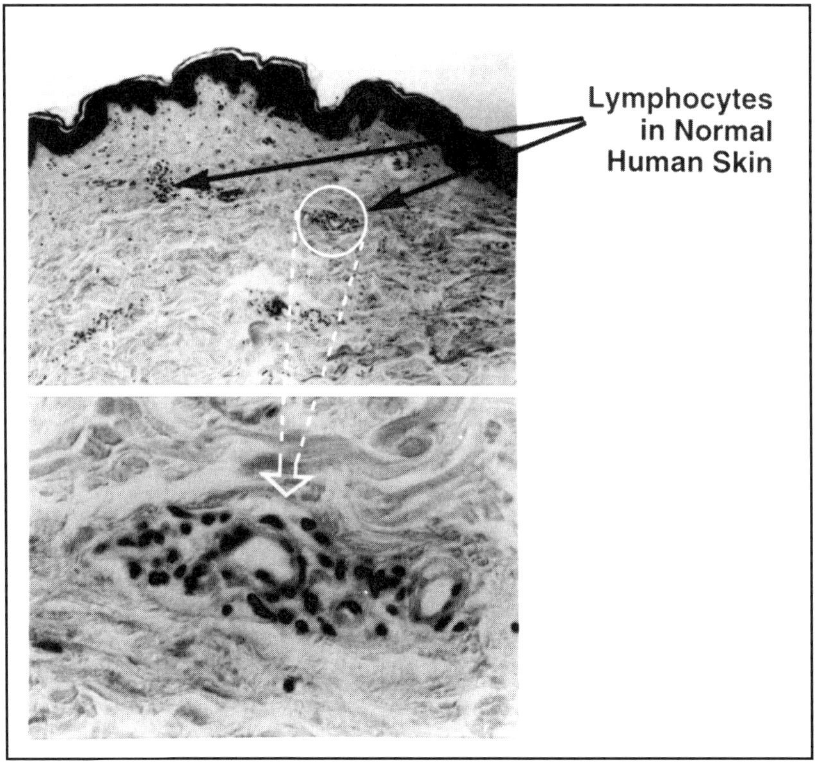

Fig. 1.4. Localization of lymphocytes in normal human skin. (a) Light microscopy of full-thickness skin showing the large distribution of lymphocytes. X100 in the original magnification. These immune cells were mainly clustered around postcapillary venules (b). X300 in the original magnification.

skin T cells are $CD45RO^+$. Functionally, cutaneous T cells may maintain a continuous state of immune surveillance and homeostasis.[49]

FUNCTIONS OF THE SKIN

The Skin as a Protective Organ

The skin is the tissue most exposed to environmental mechanical stress and pressure from differential growth of underlying organs. The skin is composed of numerous highly specialized cells actively engaged in protective functions.[31,50] Mammalian skin acts as a vital defensive structure, separating all inner organs and tissue from external environmental stress (Fig. 1.5). By its resistance, flexibility and semi-permeability, the skin prevents the whole body from mechanical injury, physiological fluid loss, microbial invasion, thermal stimuli, solar irradiation and serves as a water storage organ.

The Skin as an Immunologic Organ

All pluricellular (mammalian) organisms require an intact and functional immune system to intercept and eliminate pathogens and discard cells that have undergone malignant transformation. Anatomically, the first barrier to micro-organism invasion is the skin. However, when this physical barrier is breached, the skin uses its second arm which is immunosurveillance. For many years, it was believed that skin was a target for immune-mediated damages but had no direct role in immune function. However, the passive role of skin in immune events is no longer a tenable theory. Over the last 20 years, it has become clear that skin represents the peripheral arm of the immune system.[51,52] Also, the number of cutaneous

Table 1.6. Immunologically and nonimmunologically active cells in the skin

Cutaneous Cells	Immune Function
Keratinocytes	Yes
T lymphocytes	Yes
Langerhans cells	Yes
Melanocytes	presently No
Dendritic cells	Yes
Merkel cells	presently No
Fibroblasts	Yes*
Endothelial cells	Yes
Macrophages	Yes
Monocytes	Yes
Mast cells	Yes

* express MHC class-II after stimulation and synthesize IL-2

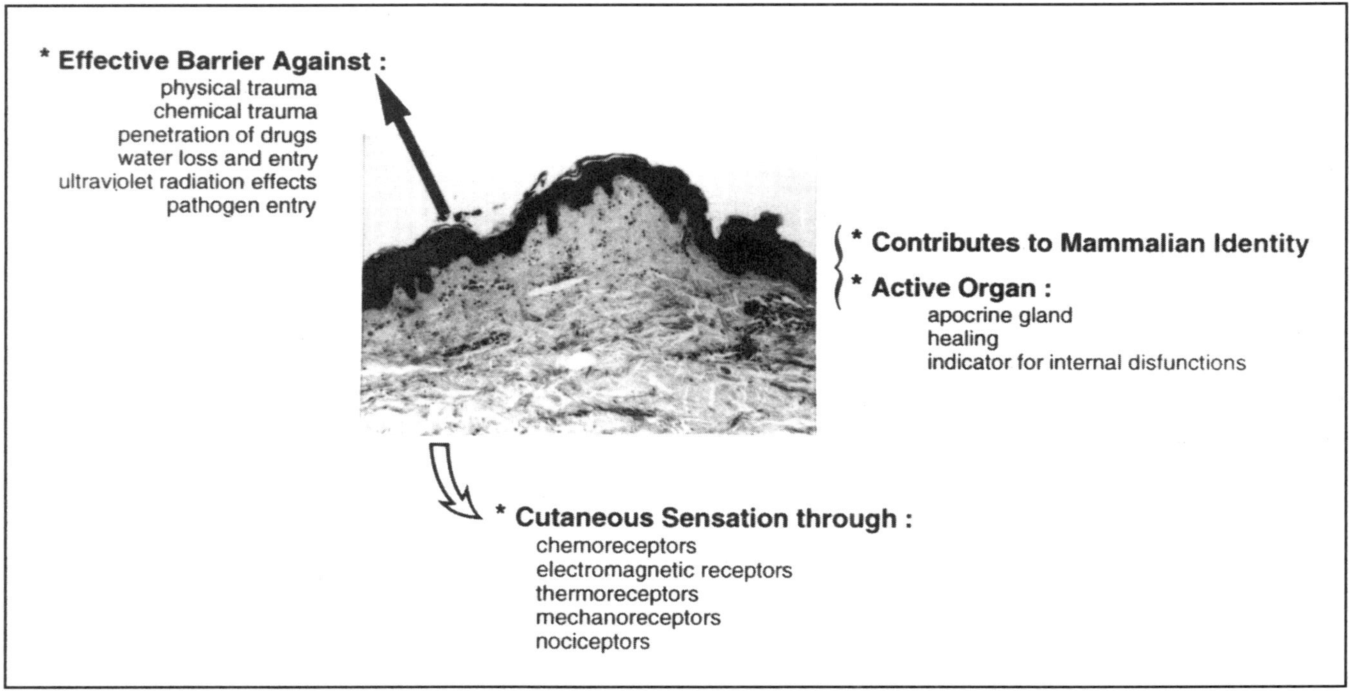

Fig. 1.5. The skin as a physiological entity for identifying and protecting the body. Each part of the skin has specific functions.

cell types known to be involved in the skin immune function is continuously increasing (Table 1.6). Consequently, only a few cutaneous cell types are still not identified as being involved in the skin's immune system. The most recent immune actor that has joined the panoply of cutaneous immune cells is keratinocyte.[53]

Keratinocyte as an Immune Cell

In the past decade, it has become clear that keratinocytes, which comprise the major proportion of epidermal cells (>90%) and whose primary function appears to be the formation of a semi-permeable barrier, participate directly and indirectly in the skin's immune system.[53,54] Indeed, keratinocytes express MHC class-I and in some situations MHC-class-II antigens and adhesion molecules. Several recent studies suggest that keratinocytes can process certain antigens and provide costimulation through cell contact. Keratinocyte-activated T lymphocytes preferentially produce T helper 2 (Th2) cytokines. By itself, keratinocytes secrete and synthesize a large number of cytokines (Fig. 1.6), and thymopoietin. This thymic hormone seems to influence the maturation of cutaneous T cells just as it does in the thymus gland. Furthermore, the skin appears to contain nurse cells with a thymocyte maturing function as those in the thymus. Keratinocytes are an important and integral part of the skin's immune system. The presence of different cells with different immune functions in the skin suggests an intra-skin communication among all these cutaneous cells and an inter-organ communication among skin, thymus and bone marrow.[51,54] However, there are still several questions which remain to be answered, such as:

A. The presence of thymopoietin and nurse cells in the skin suggests thymocyte precursor maturation in this organ. The question is whether the presence of thymopoietin in the skin is due to a local secretion or to its production in the thymus and its storage in the skin?
B. After maturation in the skin, do lymphocytes reside in cutaneous tissue or migrate through blood circulation up to the lymph nodes? Finally, do all T lymphocytes found in the skin mature in this organ?
C. It is well known that lymphocyte maturation is completed by antigen stimulation. It is also well known that the skin is the only organ which is usually in contact with various environmental

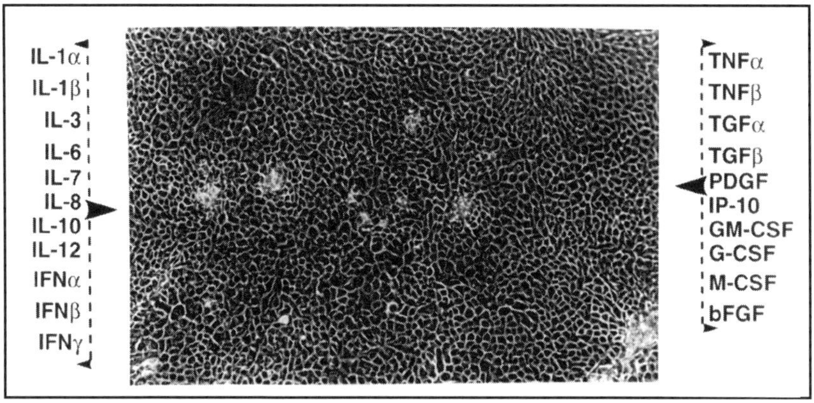

Fig. 1.6. Human keratinocytes at the confluence stage. Note the polygonal organization of these cells. In the left and the right side of the photo, different cytokines and growth factors secreted and synthesized by these epidermal cells are noted. X150 in the original magnification.

stimuli. Based on these two facts, different questions emerge: In the case of cutaneous disease, are mature T cells in the skin more active than those that mature in the thymus? Are they mobilized faster to stop disease than those maturing in the thymus?

CONCLUSION

The skin is the largest organ of the mammalian organism. It offers physical protection against hostile environments, adapts to changing conditions, heals after injury. When this barrier is breached, the skin immunologically protects the organism. Although readily accessible for study, many of the skin's functions are only now being approached by scientists. Indeed, different structures of the skin are promptly engineered and used for clinical, dermatological, cosmetological research. Different chapters will be developed in this book pointing out skin substitute production and their use, both clinically and as an experimental tool.

References

1. Rushmer RF, Buettner KJK, Short JM et al. The skin. Science 1966; 154:343-348.
2. Holbrook KA, Odland GF. Regional differences in the thickness (cell layers) of the human stratum corneum: An ultrastructural analysis. J Invest Dermatol 1974; 62:415-422.

3. Halprin KM. Epidermal 'turnover time'—a re-examination. Br J Dermatol 1972; 86:14-19.
4. Bergstresser PR, Chapman SL. Maturation of normal human epidermis without an ordered structure. Br J Dermatol 1980; 102:641-648.
5. Odland GF, Reed TH. Epidermis. In: Zelickson AS, ed. Ultrastructure of Normal and Abnormal Skin. Philadelphia: Lea & Febiger, 1967:54-75.
6. Caputo R, Peluchetti D. The junctions of normal human epidermis. A freeze-fracture study. J Ultrastruct Res 1977; 61:44-61.
7. Tseng SCG, Jarvinen MJ, Nelson WG et al. Correlation of specific keratins with different tyypes of epithelial differentiation: monoclonal antibody studies. Cell 1982; 30:361-372.
8. Elias PM, Menon GK. Structural and lipid biochemical correlates of the epidermal permeability barrier. Adv Lip Res 1991; 24:1-26.
9. Elias PM. Epidermal lipids, barrier function, and desquamation. J Invest Dermatol 1983; 80:44s-49s.
10. Dale BA, Holbrook KA, Steinert PM. Assembly of stratum corneum basic protein and keratin filaments in macrofibrils. Nature 1978; 276:729-731.
11. Sybert VP, Dale BA, Holbrook KA. Ichthyosis vulgaris: identification of a defect in synthesis of filaggrin correlated with an absence of keratohyaline granules. J Invest Dermatol 1985; 84:191-194.
12. Dale BA, Holbrook KA, Kimball JR et al. Expression of epidermal keratins and filaggrin during human fetal skin development. J Cell Biol 1985; 101:1257-1269.
13. Eckert RL, Yaffe MB, Crish JF et al. Involucrin: Structure and role in envelope assembly. J Invest Dermatol 1993; 100:613-617.
14. Brody I. The ultrastructure of the tonofibrils in the keratinization process of normal human epidermis. J Ultrastruct Res 1960; 4:264-297.
15. Wolff K, Wolff-Schreiner E. Trends in electron microscopy of skin. J Invest Dermatol 1976; 67:39-57.
16. Wolff K, Konrad K. Phagocytosis of latex beads by epidermal keratinocytes in vivo. J Ultrastruct Res 1972; 39:262-271.
17. Freinkel RK, Traczyk TN. Acid hydrolases of the epidermis: subcellular localization and relationship to cornification. J Invest Dermatol 1983; 80:441-446.
18. Bowden PE, Stark HJ, Breitkreutz D et al. Expression and modification of keratins during terminal differentiation of mammalian epidrmis. Curr Top Dev Biol 1987; 22:35-68.
19. Lavker RM, Sun T-T. Heterogeneity in epidermal basal keratinocytes: morphological and functional correlations. Science 1982; 15: 1239-1241.
20. Lavker RM, Sun T-T. Epidermal stem cells. J Invest Dermatol 1983; 81:121-127.
21. Briggaman RA, Wheeler Jr CE. Formation and origin of basal lamina and anchoring fibrils in adult human skin. J Cell Biol 1971; 51:384-395.

22. Smith LT, Holbrook KA, Byers HP. Structure of the dermal matrix during development and in the adult. J Invest Dermatol 1982; 79:93s-104s.
23. Weinstein GD, Boucek RJ. Collagen and elastin of human dermis. J Invest Dermatol 1960; 35:227-229.
24. Holbrook KA, Byers PH. Skin is a window on heritable disorders of connective tissue. Am J Med Genet 1989; 34:105-121.
25. Cotta-Pereira G, Rodrigo FG, Bittencourt-Sampaio S. Oxytalan, elaunin, and elastic fibers in the human skin. J Invest Dermatol 1976; 66:143-148.
26. Braverman IM. Ultrastructure and organization of the cutaneous microvasculature in normal and pathologic states. J Invest Dermatol 1989; 93:2s-9s.
27. Ryan TJ. Structure, pattern and shape of the blood vessels of the skin. In: Jarrett A, ed. The Physiology and Pathophysiology of the Skin. New York: Academic Press, 1973:577-805.
28. Sinclair D. Nerves of the skin: normal anatomy of sensory nerves and receptors. In: Jarrett A, ed. The Physiology and Pathophysiology of the Skin. New York: Academic Press, 1973a:348-402.
29. Ryan TJ. Structure and function of lymphatics. J Invest Dermatol 1989; 93:18s-24s.
30. Holbrook KA, Dale BA, Smith LT et al. Markers of adult skin expressed in the skin of the first trimester fetus. Curr Prob Dermatol 1987; 16:94-108.
31. Roop D. Defects in the barrier. Science 1995; 267:474-475.
32. Holbrook KA, Underwood RA, Vogel AM et al. The appearance, density, and distribution of melanocytes in human embryonic and fetal skin revealed by the anti-melanoma monoclonal antibody, HMB-45. Anatol Embryol 1989; 180:443-455.
33. Szabo G. Gerald BA, Pathak AM et al. Racial differences in the fate of melanosomes in human epidermis. Nature 1969; 222:1081-1082.
34. Wolff K, Stingl G. The Langerhans cell. J Invest Dermatol 1983; 80(6):17s-21s.
35. Romani N, Lenz A, Glassel H. Cultured human Langerhans cells resemble lymphoid dendritic cells in phenotype and function. J Invest Dermatol 1989; 93:600-609.
36. Bos JD, Kapsenberg ML. The skin immune system. Immunology Today 1986; 7(7,8):235-240.
37. Kämpgen E, Romani N, Koch F et al. Cytokine receptors on epidermal Langerhans cells. In: Moll H, ed. The Immune Functions of Epidermal Langerhans Cells. Austin: R.G. Landes Co., 1995:37-56.
38. Tachibana T. The Merkel Cell: Recent findings and unresolved problems. Arch Histol Cytol 1995; 58(4):379-396.
39. Kleinman HK, Klebe JR, Martin RG. Role of collagenous matrices in the adhesion and growth of cells. J Cell Biol 1981; 88:473-485.

40. Fradette J, Godbout M-J, Michel M et al. Localization of Merkel cells at hairless and hairy human skin sites using keratin 18. Biochem Cell Biol 1995; 73:635-639.
41. Kao W-KK, Gay S, Bressan GM et al. Demonstration by immunofluorescence that the cells from chick embryo aortas synthesize both elastin and collagen types I and III. Artery 1980; 7:176-190.
42. Weiss JB. Dermal proteins and their degradative enzymes. In: Priestley GC, ed. Molecular Aspects of Dermatology. West Sussex UK: John Wiley & Sons, 1993:111-128.
43. Beller DI, Unanue ER. Regulation of macrophage populations. II. Synthesis and expression of Ia antigens by peritoneal exudate macrophages is a transient event. J Immunol 1981; 126:263-269.
44. Tharp MD, Kagey-Sobotka A, Fox CC et al. Functional heterogeneity of human mast cells from different anatomical sites. In vitro responses to morphine sulfate. J Allergy Clin Immunol 1987b; 76:646-653.
45. Tharp MD. The mast cell and its mediator. In: Goldsmith LA, ed. Physiology, Biochemistry, and Molecular Biology of the Skin. 2nd ed. New York: Oxford University Press, 1991:1099-1120.
46. Sontheimer RD, Matsubara T, Seelig Jr LL. A macrophage phenotype for a constitutive, class II antigen-expressing, human dermal perivascular dendritic cell. J Invest Dermatol 1989; 93:154-159.
47. Larsen CP, Steinman RM, Witmer-Pack M et al. Austyn: Migration and maturation of Langerhans cells in skin transplants and explants. J Exp Med 1990; 172:1483-1493.
48. Rouabhia M, Jobin N, Doucet R et al. $CD36^+$ -dendritic epidermal cells: A putative actor in the cutaneous immune system. Cell Transplantation 1994; 3(6):529-536.
49. Bos JD, Kapsenberg ML. Lymphocyte subpopulations of the skin immune system. In: Bos JD, ed. Skin Immune System. Boca Raton: CRC Press, Inc.,1991:89-108.
50. Marks R. The skin as an organ of protection. In: Soter NA, Baden HP, eds. Pathophysiology of Dermatologic Diseases. 2nd ed. New York: McGraw-Hill, Inc., 1991:57-65.
51. Aiba S, Nakagawa S, Hara M et al. Cultured murine dermal cells can function like thymic nurse cells. J Invest Dermatol 1994; 103:162-167.
52. Bos JD, Kapsenberg ML. The skin immune system: progress in cutaneous biology. Immunology Today 1993; 14(2):75-78.
53. Nickoloff BJ, Turka LA. Immunological functions of non-professional antigen-presenting cells: new insights from studies of T-cell interactions with keratinocytes. Immunology Today 1994; 15(10):464-469.
54. Takashima A, Matsue H, Bergstresser PR et al. Interleukin-7-dependent interaction of dendritic epidermal T cells with keratinocytes. J Invest Dermatol 1995; 105(1):50s-53s.

CHAPTER 2

Exhaustive Review of Clinical Alternatives for Damaged Skin Replacement

François Berthod and Mahmoud Rouabhia

INTRODUCTION

Neglected for a long time, skin only recently is regarded as important as the heart and brain. Indeed, skin is a complex organ that serves to maintain the body's water content, protect against microorganisms and mechanical injuries, control body temperature and blood pressure. But most of all, skin is the vehicle of touch and sensation and the major repository of beauty which gives it a very special dimension. Therefore, the restoration of functionality is not the only requirement in skin replacement, good aesthetic quality, correct color and suppleness are also important. Moreover, while a partial or total destruction of an organ such as kidney, lung or liver does not necessarily impair the short-term survival of the patient, only 30% damage to the total body surface of skin can cause death. Indeed, deep burns cause dramatic fluid escape and loss of temperature control, drastic disturbances of ion equilibrium, bacterial invasion, immunodepression and pain. In addition, the disfiguring effects of burns may leave major psychological effects on a patient after inadequate healing. Thus, this severest of traumas that was the most terrifying of medieval tortures has led to extensive research in the past in order to reduce the delay of wound coverage and allow adequate healing with less pain while increasing the functional and aesthetic quality of the

Skin Substitute Production by Tissue Engineering: Clinical and Fundamental Applications, edited by Mahmoud Rouabhia.

generated skin. Although major improvements have been made over the years, much more remains to be done.

Wound Management from Antiquity to the Early 20th Century

The treatment of wounds is an art as old as humanity. Indeed, wound care was described in an Egyptian papyrus from the Pyramid Age, about 3000 to 2500 B.C.[1] Egyptian physicians stitched the lips of the wounds and treated them afterwards with honey, which served as an antiseptic, long before Fleming's discovery of penicillin.[2] Moreover, they spoke of draining suppuration from wounds, and thus stated the basis of aseptic techniques for wound management, even if their prescriptions were embellished by several incantations before treatment was begun. The Father of Medicine, Hippocrates, gave fundamental rules for wound treatment. He was the first to understand the critical importance of cleanliness and asepsis of wounds and used clean or boiled water to wash them.[3] Then, from Herophilus' practice of human vivisection to the treatment of wounded gladiators by Galen, the understanding of wound care increased through antiquity.[4] This knowledge was then lost by Europeans during the upheaval following the fall of the Roman Empire, but was kept alive by the Arabs, such as Abulcasis who later taught the Europeans.[4] Later, the French surgeon Ambroise Paré avoided the barbarous practices of the Middle Ages and promoted the systematic cleaning of wounds, following the rules written by Hippocrates seventeen centuries before.[4] In addition, he developed the concepts of pressure, positioning and splinting in the management of hypertrophic scars.[5] In the nineteenth century, a great improvement in the surgical treatment of burns was achieved with the introduction of the skin autograft technique.[6] The discovery of antibiotics by Alexander Fleming,[2] a breakthrough that led to far better wound healing and much lower morbidity, and the improvements in transplantation surgical techniques and perfusion of deep wounds, introduced by Alexis Carrel,[7] both resulted in further advances in wound management.

Over the past 40 years, tremendous progress has been made in burn treatment, leading to an improved rate of survival and rehabilitation of patients.[5] One of the major improvements has been the creation of specialized burn units, instead of burn care in general surgical departments, where patients had been managed with inadequate and conservative treatments and located in the most remote corners because of the fetid odor of decay that arose from their wounds.[8] The first burn unit may have been developed in Dzhandelidze, Russia, in 1938 and soon after World War II several burn units were created in Europe, the United States and Canada.[9,10] World War II and the Cocoanut Grove nightclub disaster in Boston in 1942, where 491 people were killed in a fire,[11] led to a number of important advances in burn treatment, which, however, became common practice only 20 years later.[8] These advances consisted mainly of treating initial burn shock and pulmonary injury, providing suitable resuscitation, limiting infections, performing early excision of burn tissues and surgically reducing scars. Indeed, the main cause of death among burn shock victims in the 1940s and 1950s was inadequate fluid resuscitation in the first hours after injury. In the 1960s and 1970s, improvements in fluid resuscitation eliminated burn shock as a major cause of mortality, while wound sepsis became the leading cause of death among burn patients. Treatment of patients with topical and systemic antimicrobial agents has since changed the principal cause of death from wound sepsis to pulmonary sepsis, following inhalation injury.[12]

Before 1950, half of the patients with a 50% total body surface burn area died.[13] Since the early 1990s, an equivalent mortality rate is only seen in patients with more than 95% of total body burn surface.[14]

Major Types of Burns

Burns are relatively common injuries that afflict millions of people in the world each year (more than one million people annually in the United States).[5] Five to ten percent of these patients require admission to a specialized burn unit and at least one percent of them die each year.[5] Major causes of burns are heat, chemicals, electricity, or radiation (Fig. 2.1).

Heat

Scalds from hot liquids and steam and contact with heating elements are the most common causes of burns, particularly in cases involving children.[15] They usually induce partial-thickness injuries, but can result in full-thickness burns depending on duration of exposure or in cases of immersion or contact with superheated vapors. Explosion of flammable liquids, gases or ammunition may induce a flash burn, characterized by a short-term exposure. This type of injury usually results in superficial burns of skin and eyes, but may cause ignition of clothing or flammable material, thus producing flame burns. Flame burns are the most serious type of heat injuries. They

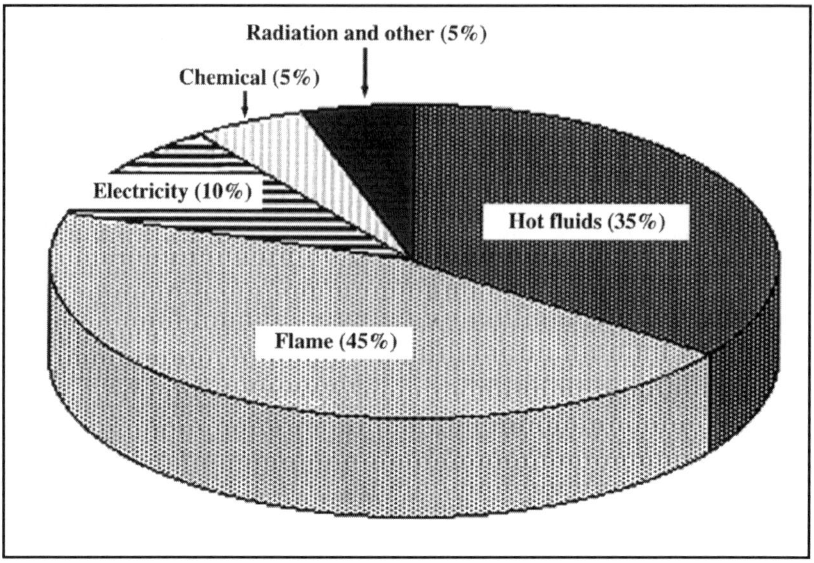

Fig. 2.1. Major causes of burns. Note that these proportions may vary depending upon country, life style, educational level and age. For example, the major cause of burn for children is scalding by hot fluids, while it is usually flame for adults. In addition, hot fluids are a major cause of burns in countries where hot tea is drunk using a conventional tea maker, instead of an automatic coffee machine.[73,74]

most often result in full-thickness and extensive burns, especially if clothing is ignited. Moreover, lung injury may occur in the case of hot smoke inhalation. This type of injury is also induced by fire immolation or gas inhalation (followed by accidental explosion) by people who attempt suicide. This corresponds to a high proportion of extensive burn patients.

It should also be mentioned that motorcycle accidents often result in deep burns induced by dermoabrasion, which are usually of small surfaces and associated with additional traumas.

Electricity

Electrical burns are due to contact with either low- (less than 1000 volts) or high-voltage electricity (more than 1000 volts).[16] Low voltage current produces a direct burn at the point of contact, and locally damages skin, subcutaneous tissues and occasionally muscles and bones.[17] High voltage current not only exposes the contact area to extremely high temperatures (2,200 to 2,800°C) for a very short duration (less than a second), but also damages tissues that conduct electricity through the body. This results in deep-tissue injury along the path of the current, that may reach internal organs such as muscles and nerves, and extensive skin burns from the combined effects of flash and flame injuries. In body parts with small cross-sectional areas, such as arms or legs, current density is high and tissue destruction severe, although in regions of large cross-sectional areas such as the trunk, current density is reduced and only mild tissue damage is observed. Burn injuries are typically characterized by extensive deep tissue damage far out of proportion to the visible skin burn.[18]

Chemicals

Chemical burns are caused by contact with acids and alkalis, organic solvents, magnesium and phosphorus. Chemicals destroy tissues by coagulation necrosis and colliquation (protein precipitation). They often result in full-thickness injuries limited to the area of contact, and may cause systemic effects by absorption through the wound.[19,20]

Radiation

Radiation burns are caused by solar and nuclear sources. Solar burns are most often limited to a superficial injury, which can involve a large body surface. Nuclear radiation exposure results in both thermal and radiation burns and may cause full-thickness and extensive injuries.[15] Radiation burn may occur in patients requiring intense superficial radiation therapy (particularly for breast cancer).

Pathophysiology of Burns

The most important feature determining the fate of a burn wound is the depth of skin injury. Basically, burns are divided into superficial, partial and full-thickness loss. Burns are typically classified in four levels (Fig. 2.2): the first degree burn, the second superficial degree burn, the second deep degree burn and the third degree burn.[21] A fourth degree burn is also described in the case of muscle injury. The first degree burn is well illustrated by solar burns and corresponds to blisters occurring in suprabasal layers of epidermis. This damage is painful but heals quickly and does not induce any risk of infection since no bleeding occurs. The second superficial degree burn results in damage to most of the epidermis and part of the papillary dermis, but allows the survival of enough islets of basal epidermal cells to promote good and complete spontaneous healing. It necessitates medical care since there are risks of infection. This burn

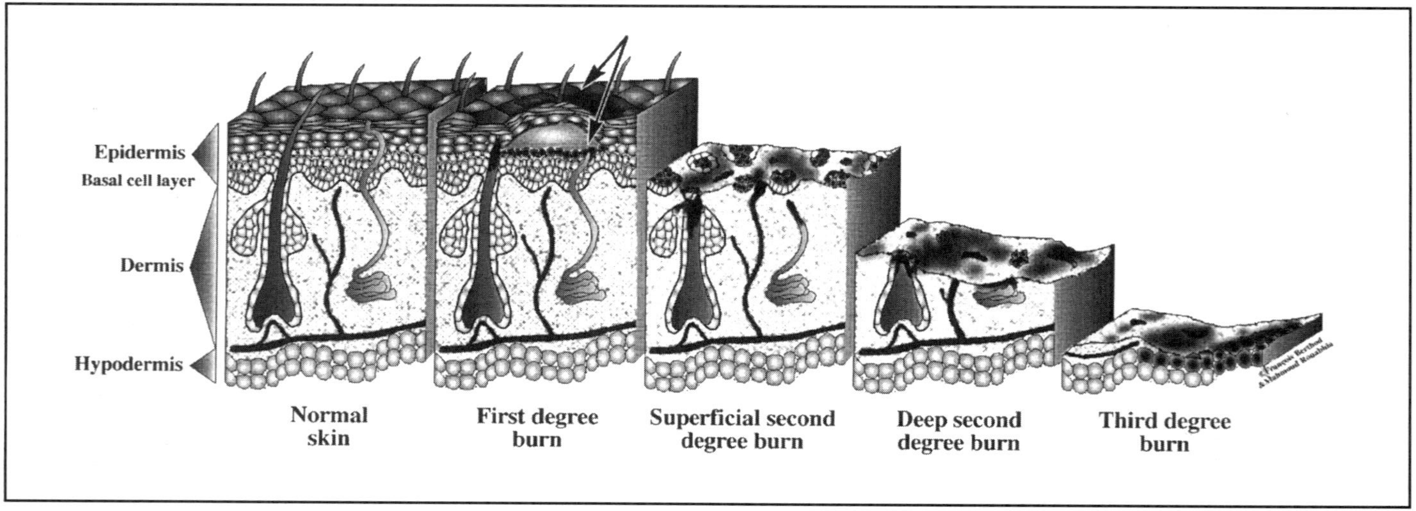

Fig. 2.2. Classification of the different levels of burns from partial to full-thickness loss. Spontaneous healing occurs in the case of first and superficial second degree burns by regeneration from epidermal basal cell layer, hair follicle and sweat gland, while deep second degree burns promote scar formation. No healing occurs in third degree burns.

Fig. 2.3 (below). Techniques of autograft expansion. Split thickness skin can be expanded with a Tanner-Vandeput mesh apparatus to obtain 1:2.5 (or more) meshed skin. It can also be expanded using the Meek island technique in order to obtain 1:4 (or more) microskin islands. Since widely meshed skin (more than 1:6) is difficult to handle, large expansion of microskin islands processed in prefolded gauze is often preferred. Widely expanded autografts are usually overlayed with homografts.

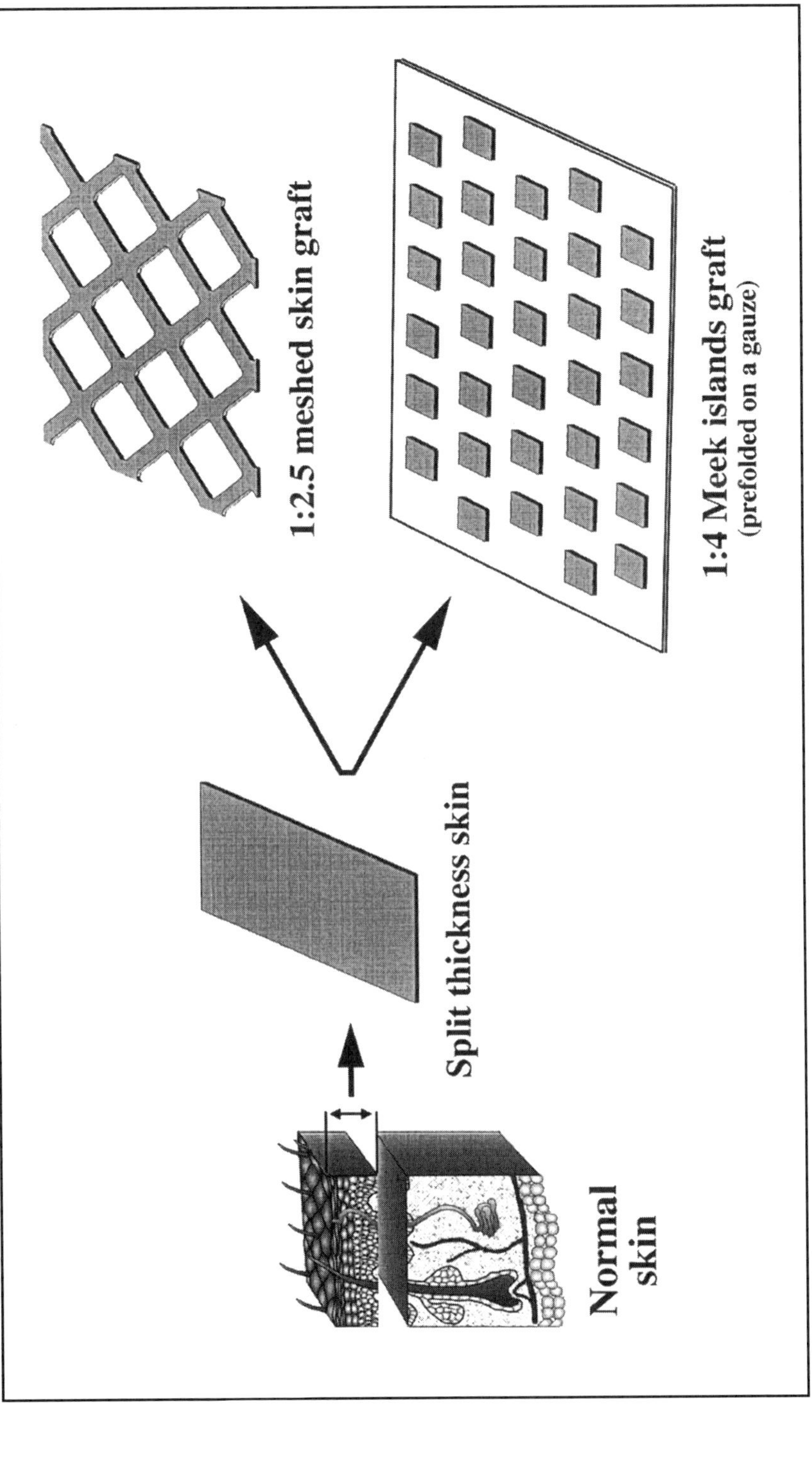
Normal skin
Split thickness skin
1:2.5 meshed skin graft
1:4 Meek islands graft
(prefolded on a gauze)

degree typically corresponds to the damage inflicted to a donor site by the taking of thin autografts (see Fig. 2.3). In the second deep degree burn, only the deeper part of the hair follicles and sebaceous and sweat glands survive with destruction of a substantial portion of the dermis. Spontaneous healing may occur from the epithelial cell source located in hair follicles, but leads to severe hypertrophic scarring and contractures, and should thus be avoided. The third degree burn involves destruction of all epidermal and dermal tissues, with occasional damage to the hypodermis, and rules out any possibility of spontaneous healing.[15]

Burns induce notable pathologic changes to the skin such as denaturation of proteins and subsequent coagulation of blood vessels leading to cell death. Three distinct areas of change in burn tissue were identified: First, the zone of coagulation, which is closest to the heat source. It essentially contains thrombosed blood vessels. Second, the zone of stasis, in which potential reversible cellular damage and intravascular sludging occur. Third, the zone of hyperemia, in which the tissue shows only a temporary inflammation with increased circulation. All burn injuries are combinations of the three zones.[22] In addition, burns induce increased capillary permeability that leads to the accumulation of large amounts of plasma in the interstitial space during the first hours after injury. This results in edema of the burn and adjacent areas, that decreases blood perfusion, thus increasing tissue necrosis.[23] It can also lead to hypovolemic shock due to loss of water and macromolecules. Burn tissues are made of a mixture of dried blood, purulent exudates produced by neutrophils and macrophages and denatured proteins of the skin. Burn wound eschars are an excellent substratum for microorganisms and become colonized by endogenous bacteria originating from the hair follicles and sweat glands.[24] Infection significantly decreases the delay of eschar detachment because it attracts leukocytes, which produce lytic enzymes that separate eschar from the underlying living tissue by digestion, but can also deepen the burn by invasive wound sepsis, and in severe cases, compromise the survival of the patient.[25]

PREPARATION OF THE BURN WOUND

Methods for burn therapy have changed in recent decades. Traditional burn management involved open and closed methods. In the open method the burns were left in a warm dry environment to crust over, while in the closed method, wounds were covered with dressings and topical treatment with antibiotics twice a day until the eschar separated. Twenty years ago, Janzekovic demonstrated that the early removal of the burn tissue reduced pain, the number of surgical procedures and the length of hospital stay. Moreover, this technique achieved better functional and aesthetic results and reduced the mortality rate.[5,26] Surgical intervention basically involves early debridement of nonviable tissue followed by burn wound closure with autografts or temporary dressings.[27] Surgical debridement of the burn wound can be performed by tangential excision,[12] which corresponds to tangentially excising slices of epidermis and dermis until viable tissue is reached as observed by arteriolar bleeding. The major drawback of this technique is the blood loss which limits the amount of burn tissue that can be removed at one time (usually no more than 20% of the total body surface).[28] Full thickness burns can also be excised down to the fat or to the fascia, particularly if the wound is infected. Excision to the fat induces the removal of the subdermal plexus of blood vessels and decreases the take of autografts since this tissue is less vascularized. Excision down to the fascia induces better autograft take but is less esthetically acceptable because of the large thickness of the removed tissues. However, this method remains useful for lifesaving treatment.

Debridement can be also achieved by enzyme digestion of the dead tissues. Proteolytic enzymes hydrolyze denatured proteins found in the eschar such as collagens, elastin, or fibrin. Debriding agents such as Travase[29] or collagenase ointment,[30] usually combined with antimicrobial agents, were used with varying degrees of success.[28] These proteolytic enzymes allow a more specific destruction of necrotic tissues while preserving viable dermis and avoiding blood loss, but the treatment is painful and can increase the risk of sepsis.[28] In addition, it takes between 6 to 10 days to achieve a clean wound bed.[30]

SKIN REPLACEMENT USING AUTOLOGOUS GRAFTS

Wound Coverage with Meshed Skin Autografts

As has been known for a long time, the best way to cover burns is the transplantation of the patient's own cutaneous tissue from an adjacent undamaged area of skin that matches closely in terms of texture, color and thickness. Split-thickness skin grafts are harvested on healthy areas using a dermatome, ranging from 0.2 to 0.5 mm deep. This surgical procedure inflicts an injury on the donor site analogous to a superficial second degree burn, allowing spontaneous healing in 2 to 3 weeks without scarring. Autograft is the method of choice to achieve definitive coverage of burned skin with good quality of healed skin. This technique has been improved since 1964 by Tanner and associates[31] by expanding the surface of the skin graft with a Tanner-Vandeput mesh apparatus. Autograft may be expanded up to 9 times its initial surface (mesh 1:9), depending on the needs of the patient (Fig. 2.3). More recently, a 25 to 30 times expansion has been reported.[32] However, when autografts of more than 1 to 3 times expansion are required, coverage of transplanted skin with $1^1/_2$ meshed homograft is necessary to protect interstices until epithelialization has been achieved. Both graft materials initially adhere to the wound; after 3 to 30 days (usually 14 days), the allograft is rejected while the autograft continues to spread to cover the site.[33] But excessive meshing usually results in healed skin which is more susceptible to infections and has a basket-like pattern which is a major drawback for aesthetic appearance.

Wound Coverage with Meek Island Autografts

An alternative expansion technique was developed by Meek in 1958.[34] The so-called Meek island graft or sandwich-graft is achieved by spray-gluing small autograft pieces, cut with a dermatome into 3 x 3 millimeters squares, onto special prefolded gauze (Fig. 2.3). When the gauze is removed after 6 days, small autograft islands are regularly spread on the wound and covered by an allograft overlay.[34,35] This method allows easier handling of widely expanded autografts than meshed skin. In addition, since the autograft islands are not mutually connected, failure of a few of them does not affect the overall graft take. The Meek technique has been reported[35] to be superior to the mesh procedure for expansion ratios of more than 1:6.

Wound Coverage with Cultured Epidermal Sheets

In the last 20 years, major improvements in resuscitation techniques allowed the survival of patients with more than 90% of total body surface burn. Thus arose the problem of covering such large surface burns that were lethal only a few years before. In such patients, early closure of burn wounds with autologous skin grafts is limited by the lack of adequate donor sites. A delay of 2 to 3 weeks is necessary to wait for healing of donor sites before harvesting them again. The initially healthy skin can usually be reharvested 2 to 3 times, as well as superficial burns after their spontaneous healing in less than 3 weeks and healed autografted wounds. But this coverage process is time-consuming and, thus, induces high risks of morbidity and mortality, mainly due to bacterial invasion. Extensive work has, therefore, been performed to develop substitutes for autologous split-thickness grafts. The purpose was to produce large surfaces of autologous skin with small areas of the patient's own skin; the answer has been the cultivation of epidermal cells in vitro. The first description of epithelial cells obtained by trypsin-splitting and cultured in vitro was reported by Medawar in 1941.[36] But epithelial cell growth was not high enough to generate the large amounts of epidermis required for grafting. A crucial improvement of the technique was carried out by Rheinwald and Green in 1975, who cocultivated irradiated fibroblasts with human keratinocytes.[37] They developed a mouse 3T3 fibroblast cell lineage requiring a sublethal irradiation that avoided fibroblast growth but maintained their potential to produce growth factors. The presence of this fibroblast feeder layer facilitates epidermal cell growth.[37] In addition, there have been major improvements in the media used for keratinocyte cultivation. This technique is now widely used throughout the world[38] and consists of the digestion of a small biopsy of healthy skin in trypsin or, more recently, in thermolysin in order to isolate keratinocytes from the basal layer of the epidermis.[39] Keratinocytes are grown for graftable sheet production in plastic flasks on a fibroblast feeder layer with culture medium enriched with growth factors and fetal calf serum until they reach subconfluency. This culture procedure offers the possibility of obtaining a 1000- to 10,000-fold larger area of graftable epidermal sheet than the initial biopsy.[37]

Although the use of epidermal autografts has markedly advanced the management of extensive burns and saved lives, this technique has major limitations.

- First, growth of cultured epidermal sheets in the laboratory needs at least 3 weeks to be achieved, thus delaying the coverage of wounds.
- Second, epidermal sheets need to be grafted on a clean wound bed since they are highly sensible to bacterial infection and toxicity of residual antiseptics.[40]
- Third, the success of the treatment strongly depends on the dexterity of the laboratory and surgical teams, from the production of the sheets to their graft and care after grafting, since this material is very fragile.

- Fourth, the regeneration of the dermal compartment underneath the epidermis is a lengthy process and skin remains fragile for at least 3 years and usually blisters.[41]
- Finally, the aesthetic aspect of the healed skin is less acceptable than the one obtained with a split-thickness graft.

Exciting developments which should eliminate these drawbacks are discussed in subsequent chapters. They may greatly improve both the speed of cultured epithelium sheet production with the use of chimeric epidermal cells and skin stem cells (see chapters 5, 6 and 8), and its quality and strength, by culturing epidermis on a reconstructed dermis (see chapters 4 and 9).

Wound Coverage with Allogeneic Dermis Overlayed by Autologous Cultured Epidermal Sheets

One of the main differences between the cultured epidermal sheet and a split-thickness autograft is the lack of the dermal structure from the cultured autograftable sheets. Absence of dermis is perceived as the major cause for a lower percentage of graft take and higher fragility and blistering after epidermal sheet transplantation, compared to split-thickness autograft. A dermal component, if present, may protect the basal layer of epidermis and have a significant impact on the postgrafting biologic responses of the epithelial cells to the differentiation and wound-healing processes. After early debridement of deep and extensive burns, temporary closure of the wound is usually achieved with cadaver allograft before autografting with cultured epidermal sheets. Instead of completely removing cadaver skin before sheet transplantation, an excision of allogeneic epidermis can be performed with a dermatome in order to only maintain the allogeneic dermis on the wound.[42] Since nonliving dermis alone may not be rejected,[43] autologous cultured epidermal sheets can be grafted onto it, thus greatly enhancing healing. Indeed, cultured epidermal sheets grafted onto homograft dermis display sooner rete ridge development and anchoring fibrils regeneration, in addition to a 95% graft take.[44]

Wound Coverage with Acellular Dermal Matrix Overlayed by Split-Thickness Autografts or Autologous Cultured Epidermal Sheets

An alternative to the removing of allogeneic epidermis from cadaver skin to maintain allogeneic dermis attached on the wound is the transplantation of acellular allograft dermal matrix. This acellular dermal matrix is produced from fresh human cadaver skin by a controlled process that removes the epidermis and the cells from the dermis without altering the structure of the extracellular matrix and the basement membrane complex.[45] Transplantation of this matrix on deep burns and its coverage with thin and widely meshed autografts allows a high percentage of take of both the dermal matrix and the autograft and avoids the undesirable scarring and contracture that usually results from the wide meshed grafts. Acellular dermal matrix produced from fresh porcine skin is also developed using a similar process in order to compensate for the lack of cadaver skin availability.[46]

Transplantation of cultured epidermal sheets on such matrices have not yet been performed but should be of great interest and might achieve results comparable to wound healing obtained with homograft dermis.

Wound Coverage with Artificial Dermis Overlayed by Autologous Cultured Epidermal Sheets

In order to cover wounds with a dermal matrix to favor graft take of cultured epidermal sheets as well as prevent rejection of homo- or xenogeneic tissues, efforts have been made to develop nonimmunogeneic artificial dermal matrices. Such dermal components must promote the prompt coverage of the largest excised full-thickness wounds, control fluid loss and prevent infections.[47] Recent advances in the technology of in vitro tissue reconstruction have made it possible to approach these requirements. As previously discussed, devitalization of allogeneic or xenogeneic dermis is a first and interesting approach to produce a dermal matrix. Other approaches have been developed in the past 20 years (see chapter 9):

- the collagen gel model, made of a mixture of fibroblasts and bovine collagen[48]
- the collagen sponge based upon the production of a lyophilized collagen matrix in which fibroblasts are cultured and migrate[49] (see also chapter 4)
- the synthetic mesh composed of a Nylon or a polyglactic acid mesh on which fibroblasts are cultured.[50]

The most exciting feature of these models is their autologous characteristic, since fibroblasts used to reconstruct them are purified from the patient's own cells. These dermal matrices can be grafted onto deep wounds after early excision and may promote the reconstruction of a new dermis suitable to support the graft of autologous cultured epidermal sheets. These products are currently undergoing clinical trials and unfortunately are not yet commercially available.

Finally, the graft in a one-step surgery of an in vitro reconstructed skin with both the autologous epidermal and dermal components seems to be the method of choice for deep burn covering, rather than the usual two-step surgery (i.e., first step, transplantation of the dermal matrix and second step, overlay with the epidermal sheet).

SKIN REPLACEMENT USING ALLOGENEIC GRAFTS

The most challenging problem in the care of patients with extensive burns is the lack of available donor sites for wound closure with autologous split-thickness skin grafts. Early excision of severe burn patient necrotic tissues needs to be followed by immediate wound coverage. Since cultured epithelial sheets require at least 3 weeks to prepare for grafting, alternative wound covers need to be used. Jackson et al reported in 1954 the use of homografts from living donors as a short-term coverage.[51] Skin allograft provides a temporary barrier to microbes and prevents fluid and protein loss. Despite rejection, generally after two weeks, wounds treated in this way improve, usually become free of infection and give better cosmetic results. Allograft coverage stimulates epithelialization from the wound margins and autologous remnants.[52] In order to extend skin allograft survival, Burke et al reported in 1974 the use of azathioprine as an immunosuppressive drug, but since infection is a major risk of morbidity in burn patients, this protocol did not gain much popularity.[53] The use of cyclosporine did not give better results.[54] Rather than immunosuppressing the patient, immunomodulation of the graft was found to be a better and simpler solution to overcome early rejection. Indeed, transplantation of living homograft has several drawbacks, including problems of supply, high immunogenicity mostly due to Langerhans cells and risk of virus transmission. Knowing that devitalization of allografts reduces their antigenicity, the use of allogeneic cadaver skin as a biological dressing is now widely accepted and usually preferred to synthetic dressings. The preservation of allografts can be performed by different techniques, such as freeze-drying,[55] glutaraldehyde fixation[56] or glycerolization.[57] Cryopreservation of homografts with glycerol is the most popular method of cadaver skin processing since freeze drying is too expensive and glutaraldehyde fixation has proven less efficient. [58] Moreover, skin preservation can reduce the risk of virus transmission from skin grafting, providing time to rid the donor skin of pathogens. Indeed, incubation of cadaver skin for several hours at 37°C in glycerol displays a significant virucidal and bactericidal effect.[57]

In order to provide sufficient cadaver skin instantly accessible for the burn patient, skin banks, such as the Euro Skin Bank in Beverwijk, The Netherlands, have been well developed through the years.[59] However, allogeneic skin banking has a significantly higher cost compared to xenogeneic skin banking and biological dressings.[21]

Wound Coverage with Large Meshed Autologous Skin Overlayed by Allogeneic Skin

Widely expanded autografts are usually required to cover extensive burns. One to six (or more) meshed skin grafts are frequently unsuccessful, although 1:3 expanded autografts usually result in graft take. Thus, cryopreserved allograft (1:1.5 meshed) is often used to cover widely meshed (1:6-1:9) autografts in order to improve their take and survival. In addition, allograft overlay prevents desiccation of the remaining vulnerable tissues on the wound bed. Allograft overlay is also used to cover Meek island autografts.[35]

Wound Coverage with Allogeneic Cultured Epidermal Sheets

As previously discussed, one of the drawbacks in the use of autologous cultured epithelial sheets for the coverage of deep and extensive burns is the time required for the growth of cells in vitro. The fact that this cultured epithelium is devoid of Langerhans cells and other MHC class II expressing cells led Thivolet et al in 1986 to hypothesize that allogeneic cultured epithelia might not be rejected. They observed an absence of T cell stimulation by the allograft throughout the follow-up of the experiment and concluded that cultured epithelial allografts were tolerated by the recipient.[60] In fact, recent clinical[61] and experimental[62] work demonstrated that allogeneic cultured epithelial sheet grafts do not survive. Even when the allograft was depleted from Langerhans cell and Thy-1^+ dendritic epidermal cells, the rejection occurred on mice after 14-16 days.[62] Readily available cultured allografts transplanted on deep partial skin thickness burns induce faster healing of the wound promoted by the residual resident keratinocytes.[63] Thus, the allograft may favor the proliferation and differentiation of spontaneously regenerating epithelium.[64] Allogeneic cultured epithelial sheets can be also frozen for storage and easier transportation, without impairing their graft efficiency.[63,65]

Wound Coverage with Amniotic Membrane

Human amniotic membranes obtained from the placenta after delivery have been used for decades to cover burn wounds.[66] These membranes are readily available in large supply in major hospitals and can be prepared relatively inexpensively. They possess most of the characteristics of an ideal skin substitute: excellent adherence to the wound, very low immunogenicity, decrease of pain, bacterial control and stimulation of healing.[67,68] Moreover, a great advantage of amnion is its translucency allowing inspection of the wound. Amniotic membranes can be applied on superficial second degree burns, donor sites and deep second degree burns after early debridement. They are also useful to cover 1:3 meshed autografts and have been reported to be extremely effective in sterilizing contaminated wounds and cleaning burns of bacteria within 3 to 5 days.[68] Nevertheless, amniotic membranes have to be changed daily and need to be covered with gauze to prevent desiccation, since they display less efficacy in preventing water loss, compared to homograft or xenograft. In addition, they do not allow long term coverage and could be dissolved by the wound early.[67] Amniotic membranes can be kept refrigerated for 6 weeks, or frozen for longer storage and banking purposes.[68]

SKIN REPLACEMENT USING XENOGENEIC GRAFT

Tissues of animal origin have been used for thousands of years to cover extensive burns. While it has become evident in our century that xenograft achieves only a temporary wound coverage, its unlimited availability under well-controlled conditions still makes animal skin a favorable wound covering. Porcine skin is the most common source of xenograft because of its high similarity to human skin. The sterility is an essential claim against xenogeneic tissues transplanted on wounds. Ionizing radiation appears to be the most suitable method to guarantee this sterility and for application in mass production.[69] Another benefit of this procedure is the significant reduction of tissue immunogenicity, since Harris and Abston demonstrated that transplantation of living pigskin induces cytotoxic lymphocyte activation in the treated patients.[70] In addition, irradiation coupled with freeze-drying seems to decrease the antigenic properties of the pigskin graft and increase its potential to inhibit bacterial growth.[69] Thus, pigskin is a well-suited temporary dressing for the coverage of second degree burns, especially after early excision, and usually promotes scar-free healing with an average healing period of about 10 days. In addition, pigskin provides a suitable overlay to cover widely meshed (1:8 to 1:12) autografts.[71] Since freeze-drying and irradiation are expensive, a low cost alternative preservation technique was successfully developed using 98% glycerin as antiseptic followed by storage at room temperature for 20 to 300 days.[72]

CONCLUSION

Improvements in resuscitation, debridement techniques and wound coverage have resulted in a higher survival and healing rate of burn patients with more than 95% of total body surface injury. Progress in the processing and banking of allogeneic and xenogeneic allograft allows a better preparation of wounds and delayed rejection. Routine production and transplantation of autologous cultured epidermal sheets for the coverage of deep and extensive burns improves the speed of permanent burn coverage, thus increasing patient survival rate (Table 2.1). But the death of 50% of burn patients with more than 90% of total body surface injury is too high. In addition, while saving the life of a patient with extensive burns, the aim of the medical team is not to create monsters with disfiguring scars. Thus, additional developments need to be realized in the care of burn patients in order to increase rate of survival and decrease scar formation. Moreover, a special effort needs to be made in the development of low cost technologies, with the purpose of sharing them, especially with Third World countries.

To attain that goal, major advances have been made owing to tissue engineering in order to produce in vitro reconstructed skin to cover deep burns. The simultaneous presence of both dermis and epidermis may greatly improve the quality of healing, while new culture techniques may allow a faster coverage of extensive burns.[46-50] Some major developments of these promising technologies will be discussed in subsequent chapters, giving an exciting view of burn treatment perspectives in the next century.

Table 2.1. Characteristics of major models for skin repacement

	Non-Antigenic	Long-Term Coverage	Healing Quality	Storage	Easy To Use	Cost	Major Additional Drawbacks	Popularity	References
Autograft	+	+	++	Not needed	+	–	Only short amount available	+++	31-33
Allograft	0	0	IE +	Frozen	+/–	++	Risks of virus transmission	++	22,51-59
Xenograft	0	0	IE +	Frozen	+	+	Risks of virus transmission	+	69-72
Autologous epidermal sheet	+	+	–	0	–	++	Fragile Available after 3 weeks	+	36-41
Allogeneic epidermal sheet	0	0	IE –	+/– Frozen	–	+	Fragile	+	60-65
Amnion	+/–	0	IE +/–	Frozen	–	–	Fragile Water loss	–	66-68
Synthetic dressings	+	0	IE –	++	++	–	Daily change	++	21
In vitro reconstructed skin	+	+	+	–	+/–	++	Unavailable	NDA	47-60

Characteristics of major models for skin replacement. IE=Indirect Effect, the transplantation of an allograft may benefit the wound and thus indirectly improve healing, but this largely depends on the type of permanent coverage that will overlay it. Note that the cost is estimated per unit of surface of materials and not for the total duration of the treatment; thus, the cost of temporary coverage materials is underestimated. NDA = No Data Available.

References

1. Breasted JH. The Edwin Smith surgical papyrus. University of Chicago, Chicago, 1930.
2. Flemming A. Streptococcal meningitis treated with penicillin. Lancet 1943; 2:434-438.
3. Richards DW. Hippocrates of Ostia. JAMA 1968; 204:1049-1056.
4. Brown H. Wound healing research through the ages. In: Cohen, IK, Diegelmann, RF, Linblad WJ, eds. Wound Healing, Biochemical and Clinical Aspects. Philadelphia: WB Saunders Company, 1992:5-18.
5. Nguyen TT, Gilpin DA, Meyer NA et al. Current treatment of severely burned patients. Ann Surg 1996; 223:14-25.
6. Ollier LXEL. Sur les greffes cutanées autoplastiques. Bull Acad Med 1872; 1:243.
7. Carrel A. Carrel-Dakin solution. JAMA 1916; 67:1777-1778.
8. Monafo WW. Then and now: 50 years of burn treatment. Burns 1992; 18(S2):S7-S10.
9. Marvin JA. Burn nursing history: the history of burn care. J Burn Care Rehabil 1993; 14:252-257.
10. Dimick AR, Brigham PA, Sheehy EM. The development of burn centers in North America. J Burn Care Rehabil 1993; 14:284-299.
11. Saffle JR. The 1942 fire at Boston's Cocoanut Grove nightclub. Am J Surg 1993; 166:581-591.
12. Muller MJ, Herndon DN. The challenge of burns. Lancet 1994; 343:216-220.
13. Bull JP, Fisher AJ. A study of mortality in a burn unit: a revised estimate. Ann Surg 1954; 139:269-274.
14. Herndon DN, Rutan RL. Comparison of cultured epidermal autograft and massive excision with serial autografting plus homograft overlay. J Burn Care Rehabil 1992; 13:154-157.
15. Wachtel TL. Initial care of major burns. Postgrad Med 1989; 85:179-196.
16. Hunt JL, Sato RM, Baxter CR. Acute electric burns: current diagnostic and therapeutic approaches to management. Arch Surg 1980; 115:434-438.
17. Rouge D, Polynice A, Grolleau JL et al. Histologic assessment of low-voltage electrical burns: experimental study with pigskin. J Burn Care Rehabil 1994; 15:328-334.
18. Goodwin CW, Finkelstein JL, Madden MR. Burns. In: Schwartz SI, Shires GT, Spencer FC, eds. Principles of Surgery. 6th ed. New York: McGraw-Hill, Inc., 1994:225-277.
19. Boykin JV, Molnar JA. Burn scar and skin equivalents. In: Cohen, IK, Diegelmann RF, Linblad WJ, eds. Wound healing, Biochemical and clinical aspects. Philadelphia: W.B. Saunders Company, 1992: 523-540.
20. Erdmann D, Hussman J, Kucan JO. Treatment of severe alkali burn. Burns 1996; 22:141-146.

21. MacMillan BG. Present status of bioadherent materials, barrier dressings, and biosynthetics as skin substitutes. In: Wise DL, ed. Burn Wound Coverings.Vol 1. Boca Raton: CRC Press, 1984:115-136.
22. Jackson DM. The diagnosis of the depth of burning. Br J Surg 1953; 40:588-596.
23. Baxter CR. Fluid volume and electrolyte changes of the early postburn period. Clin Plast Surg 1974; 1:693-701.
24. Hammond JS, Ward CG. Complications of the burn injury. Crit Care Clin 1985; 1:175-187.
25. Luterman A, Dasco CC, Curreri PW. Infections in burns. Am J Med 1986; 81:45-52.
26. Janzekovic A. A new concept in the early excision and immediate grafting of the burns. J Trauma 1975; 15:42-62.
27. Tomkins RG. Significant reductions in mortality for children with burn injuries through the use of prompt eschar excision. Ann Surg 1988; 208:577-585.
28. Markepease AR. Enzymatic debridement of burns: a review. Burns 1982; 9:153-157.
29. Shakespeare PG, Harris AG. The activity of the enzymatic debridement agent Travase towards a variety of protein substrates. Burns 1979; 6:15-20.
30. Soroff HS, Sasvary DH. Collagenase ointment and polymixin B sulfate/bacitracin spray versus silver sulfadiazine cream in partial-thickness burns: a pilot study. J Burn Care Rehabil 1994; 15:13-17.
31. Tanner JC, Vandeput J, Olley JF. The mesh skin graft. Plast Reconstr Surg 1964; 34:287-292.
32. Richard R, Miller SF, Steinlage R et al. A comparison of the Tanner and Bioplasty skin mesher systems for maximal skin graft expansion. J Burn Care Rehabil 1993; 14:690-695.
33. de Riel S. Assessment of burn wound therapy systems. In: Wise DL, ed. Burn wound coverings. Vol 1. Boca Raton: CRC Press, 1984:1-37.
34. Meek CP. Successful micrografting using the Meek-Wall microdermatome. Am J Surgery 1958; 96:557-558.
35. Kreis RW, Mackie DP, Hermans RP et al. Expansion techniques for skin grafts: comparison between mesh and Meek island (sandwich) grafts. Burns 1994; 20(S1):S39-S42.
36. Medawar PB. Sheets of pure epidermal epithelium from human skin. Nature (London) 1941; 148:783-784.
37. Rheinwald JG, Green H. Serial cultivation of strains of human epidermal keratinocytes: the formation of keratinizing colonies from single cells. Cell 1975; 6:331-344.
38. Odessey R. Addendum: multicenter experience with cultured epidermal autograft for treatment of burns. J Burn Care Rehabil 1992; 13:174-180.
39. Germain L, Rouabhia M, Guignard R et al. Improvement of human keratinocyte isolation and culturing using thermolysin. Burns 1993; 19:99-104.

40. Sheridan RL, Tompkins RG. Cultured autologous epithelium in patients with burns of ninety percent or more of the body surface. J Trauma 1995; 38:48-50.
41. Putland M, Snelling CF, Macdonald I et al. Histologic comparison of cultured epithelial autograft and meshed expanded split-thickness skin graft. J Burn Care Rehabil 1995; 16:627-640.
42. Cuono C, Langdon R, McGuire J. Use of cultured epidermal autografts and dermal allografts as skin replacement after burn injury. Lancet 1986; 17:1123-1124.
43. Yukna RA, Turner DW, Robinson LJ. Variable antigenicity of lyophilized allogeneic and lyophilized xenogeneic skin in guinea pigs. J Periodont Res 1977; 12:197-201.
44. Compton CC, Hickerson W, Nadire K et al. Acceleration of skin regeneration from cultured epithelial autografts by transplantation to homograft dermis. J Burn Care Rehabil 1993; 14:653-662.
45. Wainwright DJ. Use of an acellular allograft dermal matrix (Alloderm) in the management of full-thickness burns. Burns 1995; 21:243-248.
46. Livesey SA, Herndon DN, Hollyoak MA et al. Transplanted acellular allograft dermal matrix: potential as a template for reconstruction of viable dermis. Transplantation 1995; 60:1-9.
47. Yannas IV, Burke JF, Warpehoski M et al. Prompt, long-term functional replacement of skin. In: Wise DL, ed. Burn wound coverings. Vol 2. Boca Raton: CRC Press, 1984:27-33.
48. Bell E, Erhlich HP, Buttle DJ et al. A living tissue formed in vitro and accepted as a full thickness skin equivalent. Science 1981; 211:1042-1054.
49. Berthod F, Saintigny G, Chretien F et al. Optimization of thickness, pore size and mechanical properties of a biomaterial designed for deep burn coverage. Clin Mater 1994; 15:259-265.
50. Rennekampff HO, Kiessig V, Hansbrough JF. Current concepts in the development of cultured skin replacements. J Surg Res 1996; 62:288-295.
51. Jackson D. Clinical study on the use of skin homografts for burns. Br J Plast Surg 1954; 7:26-43.
52. Wong L. The many uses of allograft skin. Ostomy Wound Manage 1995; 41:36-42.
53. Burke JF, May JW jr, Albright N et al. Temporary skin transplantation and immunosupression for extensive burns. New Engl J Med 1974; 290:269-271.
54. Eldad A, Benmeir P, Weinberg A et al. Cyclosporin A treatment failed to extend skin allograft survival in two burn patients. Burns 1994; 20:262-264.
55. Abbott WM, Hembree JS. Absence of antigenicity in freeze-dried skin allografts. Cryobiology 1970; 6:416-418.
56. Schechter I. Prolonged retention of glutaraldehyde-treated skin allografts and xenografts. Ann Surg 1975; 182:699-703.

57. Hussman J, Russel RC, Kucan JO et al. Use of glycerolized human allografts as temporary and permanent cover in adults and children. Burns 1994; 20(S1):S61-S66.
58. Hettich R, Ghofrani A, Hafeman B. The immunogenicity of glycerol-preserved donor skin. Burns 1994; 20(S1):S71-S76.
59. de Backere ACJ. Euro Skin Bank: large scale skin-banking in Europe based on glycerol-preservation of donor skin. Burns 1994; 20(S1)S4-S9.
60. Thivolet J, Faure M, Demidem A et al. Long-term survival and immunological tolerance of human epidermal allografts produced in culture. Transplantation 1986; 42:274-280.
61. Auböck J, Irschick E, Romani N et al. Rejection after a slightly prolonged survival time of Langerhans cell-free allogeneic cultured epidermis used for wound coverage in humans. Transplantation 1988; 45:730-737.
62. Rouabhia M, Germain L, Bélanger F et al. Cultured epithelium allografts: Langerhans cell and Thy-1^+ dendritic epidermal cell depletion effects on allograft rejection. Transplantation 1993; 56:259-264.
63. de Luca M, Albanese E, Bondanza S et al. Multicentre experience in the treatment of burns with autologous and allogeneic cultured epithelium, fresh or preserved in a frozen state. Burns 1989; 15:303-309.
64. Oliver AM, Kaawach W, Mithoff EW et al. The differentiation and proliferation of newly formed epidermis on wounds treated with epithelial allografts. Br J Dermatol 1991; 125:147-154.
65. Madden MR, LaBruna AA, Hajjar DP, Staiano-Coico L. Transplantation of cryopreserved cultured epidermal allografts. J Trauma 1996; 40:743-750.
66. Stern M. The grafting of preserved amniotic membrane to burned and ulcerated surfaces substituting skin grafts. A preliminary report. JAMA 1913; 60:973.
67. Thomson PD, Parks DH. Amnion as a burn wound dressing. In: Wise DL, ed. Burn Wound Coverings. Vol 1. Boca Raton: CRC Press, 1984:47-53.
68. Walker AB. Use of amniotic membranes for burn wound coverage. In: Wise DL, ed. Burn Wound Coverings. Vol 1. Boca Raton: CRC Press, 1984:55-83.
69. Pellet S, Ménesi L Novák J et al. Freeze-dried irradiated porcine skin as a burn wound covering. In: Wise DL, ed. Burn Wound Coverings. Vol 1. Boca Raton: CRC Press, 1984:85-114.
70. Harris NS, Abston S. Anti-porcine antibodies in xenografted patients. J Surg Res 1974; 16:599-600.
71. Lin SD, Lai CS, Chou CK et al. Microskin autograft with pigskin xenograft overlay: a preliminary report of studies on patients. Burns 1992; 18:321-325.

72. Basile ARD. A comparative study of glycerinized and lyophilized porcine skin in dressings for third-degree burns. Plast Reconstruct Surg 1982; 69:969-972.
73. Uçar N, Haberal M. In-depth investigation of the causes and prevention of minor burns treated at a Turkish burn centre. In: Masellis M, Gunn SWA, eds. The Management of Burns and Fire Disasters, Perspectives 2000. Boston: Kluwer Academic Publishers, 1995:139-142.
74. Golubovic Z, Janjic G, Parabucki D, Petronic I. Burns prevention in a paediatric population. In: Masellis M, Gunn SWA, eds. The Management of Burns and Fire Disasters, Perspectives 2000. Boston: Kluwer Academic Publishers, 1995:131-138.

CHAPTER 3

Cultured Autologous Epidermis for Massive Burn Wounds: 15 Years of Practice

Odile Damour, Fabienne Braye, Jean-Louis Foyatier,
Anne Fabreguette, Patricia Rousselle, Severine Vissac, Paul Petit

INTRODUCTION

Progress in the field of medical and surgical resuscitation has led to an increase in the survival rate of patients affected with extensive full thickness burns. The modern treatment of deep burns consists of early excision of burned tissue followed by immediate coverage with autograft.[1,2] However, in the case of extensive burns, healthy skin is insufficient to cover massive wounds even when the maximal expansion is obtained with mesh dermatome. Although iterative samples can be done, the definitive wound closure must be attempted rapidly in order to avoid septic complications which are the main cause of death. Moreover, when the donor sites have been harvested several times, the healing occurs slowly and after three times, skin quality is diminished. These problems emphasize the fact that there is an essential need for skin substitutes. Different approaches have been attempted for temporary coverage of the wound; these are allogenic human skin,[3] xenogenic pig skin,[4] amniotic membranes and biosynthetic materials. The functions of these temporary coverages are protection and preparation of the wound before epithelialization.

Since 1940, many teams have focused their research on storage of skin and isolation of epidermal cells.[5] In 1956, Basset[6] and Earle[7]

Skin Substitute Production by Tissue Engineering: Clinical and Fundamental Applications, edited by Mahmoud Rouabhia.

proposed that epidermal cell culture in massive quantities could provide a radical solution for wound closure in burns.

Keratinocytes were known as difficult cells to grow, because of their limited proliferation, early differentiation and frequent contamination with fibroblasts.

In 1956, Puck[8] grew epithelial cells on a feeder layer of irradiated epithelial cells. This technique was based on the fact that irradiated cells lose their growing capacity, but retain their ability to synthesize growth factors and proteins.

In 1971, Karaseck[9] demonstrated the importance of fibroblasts for epidermal growth and differentiation. Rheinwald and Green[10] were the first to obtain, in 1975, serial cultivation of human keratinocytes on a feeder layer of irradiated 3T3 fibroblasts. Then, in 1979, Green[11] presented a method to prepare cultured autologous epithelia suitable for grafting.

This technique has now become fully integrated into the therapeutic strategy of such patients. The first clinical experience with cultured autologous epithelium was reported in 1981 by O'Connor et al,[12] Gallico[13] in 1984 and then by numerous other teams in the world. Results of these experiments are compiled in Table 3.1. This chapter will present, in its first part, the results obtained by several authors, and in its second part, an overview of 10 years of clinical experience in our burn centers. We will finally discuss the different possible clinical approaches.

BIOLOGICAL ASPECTS

Preparation of the Cultured Autologous Epithelia

It is now well established that the culture of keratinocytes on feeder layers is the best technique to obtain successful and reproducible production of sufficient cultured epithelia. Other systems use defined culture medium replacing the serum with bovine pituitary extract (BPE).[14] This culture system provides cells with hyperproliferative growth potential[15,16] when the calcium concentration is low, and differentiating cells suitable for grafting[17] when the calcium concentration is high. However, since 1989, CAE on feeder layers are preferred because of the reluctance to use defined culture medium with BPE. Indeed, the pituitary extract is more likely a source of bovine spongiform encephalopathy than is fetal calf serum.[18] The feeder layer is composed of lethally irradiated fibroblasts. Pretreating fibroblasts with mitomycin is an alternative to irradiation.[19] All fibroblast cell lines do not support keratinocyte growth at the same level. However, the 3T3-J2 murine fibroblast line[20,21] and normal human fibroblasts[22] allow a similar growth rate of keratinocytes.

After adhesion on extracellular matrix components synthesized and deposited by irradiated fibroblasts (Fig. 3.1), keratinocytes isolated from skin biopsies produce three types of clones: holoclones, meroclones and paraclones. Only basal cells grow as holoclones. They are considered as stem cells and have the highest growth potential with 50 possible generations allowing, in theory, five to six subcultures. Suprabasal cells which produce paraclones have the lowest growth potential with a maximal lifetime of 15 generations.[23] During cell culture or aging, holoclones become progressively meroclones and paraclones. In that regard, and in order to avoid aging of the cells, only second or third passages are used for grafting. Each colony expands at the periphery, cells dividing and migrating outwards.

The keratinocyte suspension is seeded on the feeder layer in a medium containing fetal calf serum and several supplements. One of them, epidermal growth factor (EGF), acts when differentiation and stratification starts at the center of colonies and increases the potential of migration of keratinocytes to the periphery.[24] It follows that, culture lifetime of keratinocytes increases in culture flasks from 50 to 150 generations.[25] Cholera toxin and all of the cyclic-AMP activators increase the cellular growth rate[26] through the integrins.[27]

Within the first few days following admission of the patient, a full thickness skin biopsy (4-8 cm^2) of healthy skin is harvested as a skin graft after disinfection by a bactericide and sporicide antiseptic such as polyvidone-iodine. The sample is immediately immersed into Dubelco modified Eagle's medium (DMEM) containing antibiotics and is rapidly transported at 4°C to the laboratory. Keratinocytes are isolated after enzyme treatment, then seeded in DMEM supplemented medium. After three days of culture, 10 ng/ml of human recombinant epidermal growth factor (EGF, Austral Biology, USA) is added to the medium. When the cultures are nearly confluent, 8-10 days later, the keratinocytes are enzymatically resuspended, then seeded at 3-6 x 10^3 cells per cm^2 to establish secondary cultures. These cultures are maintained as the primary cultures and usually become confluent within 10-12 days. For grafting, secondary and sometimes tertiary

Table 3.1. Cultured autologous epithelia for burn treatment

Investigators Date	No. of Patients	Age Average (yrs)	Burns % — % Total of BSA* Burned	% Full Skin Thickness	% Graft Take
Cuono[33] (1986)	2	47 yrs 32 yrs	55% 65%	45%	
Pittelkow[17] (1987)	20				40-60%
Eldad[34] (1987)	2				14-70%
Herzog[35] (1988)	8		16-88%		0-85%
Kumagai[36] (1988)	6	28 yrs (12-46 yrs)	50% (5-75%)		2 loss of grafts 3 took partly 1 took completely
Siwy[37] (1989-1991)	5	31.6 yrs (14-50 yrs)	65% (50-85%)		75.4% (60-80%)
O'Connor[38] (1989)	18	8.8 yrs (5 mos-16 yrs)	79%		55% (12-80%)
Weekley[39] (1989)	10				72%
De Luca[40] (1989)	40	22 yrs (2-59 yrs)	53% (10-95%)	37.8% (7-90%)	37% ≤ 18 yrs 47% > 18 yrs 28%
Teepe[41] (1990)	17	30 yrs (2-56 yrs)	56% (31-85%)		15-47%
Munster[42] (1990)	7	31 yrs	69.6% (54 -91%)	31.4% (13-41%)	75%
Foyatier[43] (1990)	16	34 yrs (6-88 yrs)	66% (30-92%)	44% (20-75%)	> 50% for 9 pts < 50% for 3 pts 0% for 4 pts
Tamisani[44] (1991)	5	3.8 yrs (2-9 yrs)	25% (6-40%)	14% (7-20%)	78% (30-100%)
Haith[45] (1992)	6	40 yrs (19-69 yrs)	70% (40-99%)		78% (52-100%)
Nave[46] (1992)	3	41 yrs (28-62 yrs)	80% (70-90%)		84.5% (81 and 88%)
Barillo[47] (1992)	3	36.6 yrs (20-67 yrs)	59% (45-86%)	47% (29-76%)	62.5% (55-75%)

continued...

Table 3.1. Cultured autologous epithelia for burn treatment

Investigators Date	No. of Patients	Age Average (yrs)	Burns % — % Total of BSA* Burned	% Full Skin Thickness	% Graft Take
Donati[48] (1992)	12	27 yrs (10-80 yrs)	36% (15-75%)	31.5% (15-75%)	25.8% (0 à 50%)
	14	30.5 yrs (9-65 yrs)	30.4% (10-70%)	18.4% (10-50%)	37.5% (0-65%)
Herndon[49] (1992)	1	10 yrs	> 80%	> 80%	initial: 90% with biosurface 50% with homeculture final: 30%
De Luca[50] (1992)	14	8.6 yrs (1-18 yrs)		35.2% (7-87%)	64% (10-100%)
Krupp [51] (1992)	21	3-71yrs			75%
Mc Aree[52] (1993)	7	21 yrs (9-45 yrs)	66% (40-85%)	52% (35-74%)	initial: 60% (57-95%) final: 80% (56-98%)
Rue[53] (1993)	6	31 yrs (18-48 yrs)	53.5% (42-64%)		initial: 72% (10-85) final: 73.5% (50-88)
	10	29 yrs (10-56 yrs)	77.1% (70-85%)		initial: 60% (50-87) final: 30.6% (0-85)
Hicherson[54] (1994)	5	38 yrs (20-60 yrs)	60% (48-70%)		93% (87-100%)
Sheridan[55] (1995)	5	11.5 yrs (7 mos-37 yrs)	94% (90-96%)		initial: 51% (20-80%) final: 33% (20-50%)
Carsin[56] (1996)	19	2-70 yrs	77.7% (60-95%)	66.9% (20-90%)	72.3%
Still[57]	15	2-56 yrs	54-85%		71.5%
Herriot Burn Centre Lyon	64	28 yrs 1-75yrs	62% (22-95%)	59.8% (20-90%)	57.5% (10-95%)

* BSA, Body surface area

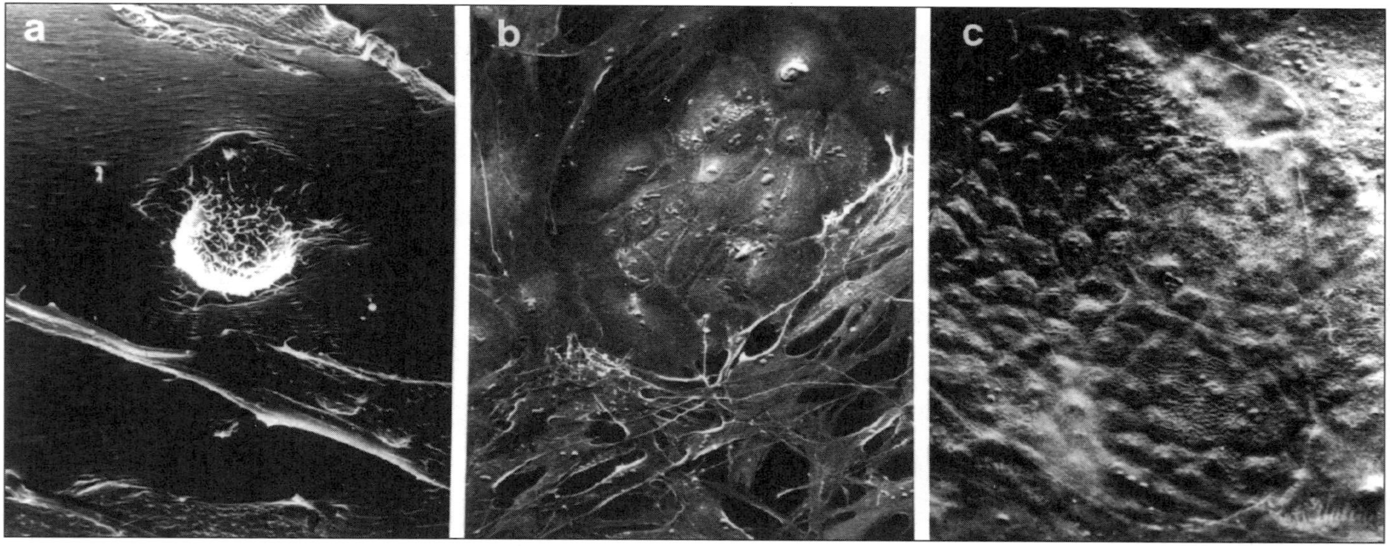

Fig. 3.1. Scanning electron microscopy of a keratinocyte culture (scale = 18.5 µm). (a) A keratinocyte adhering on extracellular matrix (ECM) synthetized by irradiated normal human fibroblasts (NHF). (b) A strain developed from a basal cell. We may observe that the keratinocytes drive off the NHF. (c) Confluent keratinocytes give a sheet ready to be detached from the bottom of the flask. Note the complete absence of fibroblasts.

cultures can be used. Confluent epithelia are enzymatically detached as a coherent sheet from the culture flask using 1.2 U/ml Dispase II (Boehringer Mannheim, Germany)[28,29] in a serum-free medium. The sheets are mounted on vaseline gauze (Sherwood) with the basal side upwards. When the cultures begin to detach from the plastic flask, the periphery of the sheet is flipped over the gauze. Finally, the sheets are clipped, transferred to petri dishes containing DMEM, and maintained under 5% CO_2 until grafting (Fig. 3.2). Other teams culture the keratinocytes directly on a support suitable for grafting such as benzyl ester of hyalurnic acid[30] on fibrin glue[31] or in fibrin glue suspension.[32]

REVIEW OF CAE CLINICAL RESULTS

Table 3.1 regroups the results obtained by different teams throughout the world using the technique based on the Rheinwald and Green method from 1981 to 1996. Our initial intention was to collect this information and correlate the take rate with various clinical criteria such as age, total body surface area (TBSA), depth and

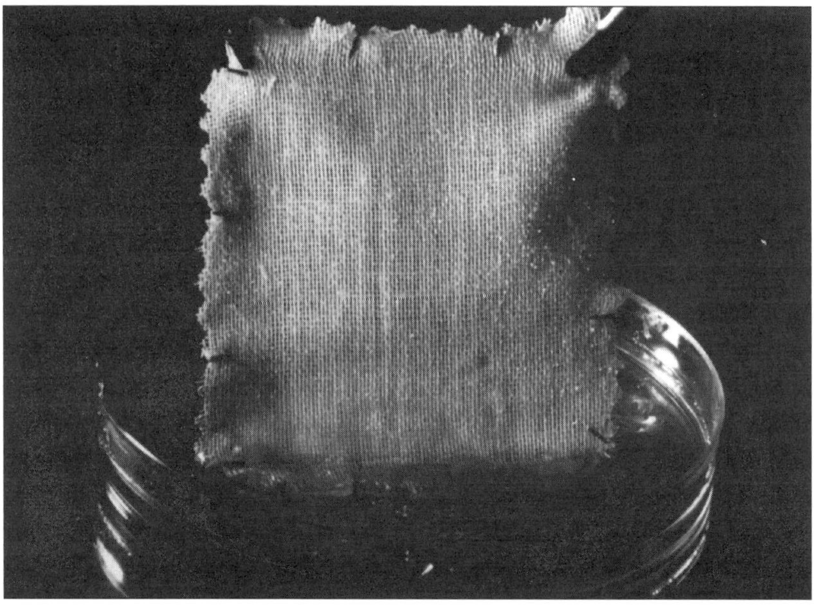

Fig. 3.2. Cultured autologous epidermis in the transfer medium ready to be grafted.

grafted surface with CAE. After taking into consideration the obvious lack of standardization in the expression of take rates, we were not able to perform any statistical analysis.

The epidermal sheets have either been prepared by a proximity culture laboratory[33-38] or were commercially available.[39-47]

Because some of the articles described identical patient's data, we have chosen 27 papers out of 32 which are presented in Table 3.1. Indeed, the results were published at different times after the graft, either as cumulated data in review articles or, for a few of them, as part of a description of a long-term patient outcome.[48-52]

This table shows that the population of treated patients is highly variable from 1 to 40 patients depending not only on length of the study, but also on the activity of the burn unit. The age of the patients is unknown in ten reports making calculation of the average impossible. The youngest patient was 5 months old and the eldest 88 years old.

The total burn area, which has been omitted for seven patients, ranged from 10% to 95%. The deeply burned surface which ranged from 7% to 90% of the burned surface appears only in nine papers. However, we may observe that, since 1990, CAE were used mostly for the treatment of over 50% burned patients. Moreover, the grafted surface by CAE is not often clearly reported and also showed a great variability ranging from 170 cm^2 to 5225 cm^2.

The number of deaths has only been documented in a few reports and we cannot answer the Heimbach question:[53] "What are the survival rates for the patients with large burns in burn centers that use CEA?"

Concerning the results, basically the take rate of CAE is lower than that observed with split-thickness autografts (75% versus 92%) and is also correlated with the metabolic state of the patient.[41,48]

The results presented in Table 3.1 demonstrate that the take rate shows a great variability ranging from 0% to 100%, but mostly comprised between 30 to 85%. In fact, we can conclude that experiments undertaken since 1990 under the best clinical conditions yielded a 50% or more take rate. However, we cannot clarify if the number of regrafting is taken into account in the evaluation of the take rate, knowing that the number of CAE applications is reported only in 10 papers with an average of 2 times, ranged between 1 to 7 times, excluding the 19-times grafted case of Haith.[42]

Again, the interpretation of the results is also made difficult by the absence of standardized evaluation criteria. Most of the teams carried out exclusively visual evaluations and follow-up of the graft. These were often performed after different post-grafting delays. It appears that the take rate is dependent on the date of evaluation. Thus, it is important to know this date because authors report a secondary loss of the coverage 2, 3 or 4 weeks after the take. It can be total or partial, with or without infection. The partial secondary loss can also be attributed to a fragility due to inadequate anchorage of the graft to the underlaying tissue.[54,55] This abnormality causes a lack of long-term graft stability.[46,56]

The exact reasons for the discrepancies are the intervention of multiparameters in the take rate, including wound bed, surgical excision, infection, nutrition, general conditions and age.[34,57] While these parameters are the same for classical autografts, some of them have a more important impact, for example, infection and handling of the grafts during post-operative wound management.[54,58]

Odessey's study[59] is of great interest since the parameters of treatment and evaluation were standardized. Of 240 patients in 79 burn centers treated with industrially provided CAE during two and a half years, Odessey was able to analyze 104 results because all the conditions were standardized. The final take rate is calculated 3 or 4 weeks after grafting as follows:

$$\frac{(\text{grafted area - grafted area requiring another replacement}) \times 100}{\text{area grafted with CAE}}$$

The average size burn treated with CAE was 70% (range, 25% to 99%), 50% of which was full-thickness. The average of final take was approximately 60% for the 104 patients and the repartition of the population in function of the take rate is presented in Table 3.2.

Table 3.2. Repartition of population as a function of final take (Odessey, 1992)

% Population (n=104)	% Take Rate
16%	< 30%
50%	30-70%
22%	70-90%

He also demonstrated, in agreement with several authors, that the most important parameter is wound preparation prior to grafting. It is also our point of view that the take rate and aesthetic and functional results are dependent on the presence of the dermis and that the autologous dermis must be preserved when possible. This is the strongest argument for saving the maximum of the dermis during the excision and for encouraging the use of combined techniques using large meshed autograft covered with CAE, autograft bringing both dermis and epidermis.[60,61] This technique will be described elsewhere. When excision to aponeurosis is necessary, the best preparation of the wound is temporary coverage with viable homograft (Fig. 3.3) when allodermis can be preserved.[38,45,58,61-67] However, this technique requires previous dermabrasion of alloepidermis. Foyatier,[36] in agreement with Sheridan,[58] underlines the technical difficulties of this method in limiting the homograft abrasion to its superficial area. The attachment portion of the dermis is superficial and may, in part, be removed with the alloepidermis decreasing the chances of CAE take success. On the other hand, the risk of viral transmission, inherent in the use of allograft, and lack of skin donor, led to the development of other solutions such as the combined technique. The use of glycerolyzed allografts provides low viral transmission risks, attenuates the antigenicity and allows repopulation with fibroblasts because they are acellular.[68-70] All these advantages do not solve the problem of lack of availability. These reasons encouraged Yannas and Burke[70-72] to develop a reconstructed dermis substrate which is now available. This product has not yet proved its efficiency as a wound bed preparation for CAE. This can be explained either by the presence of silicone recovering the collagen-glycosaminoglycans matrix or the use of glutaraldehyde as a cross-linker of collagen. In the mean time, it was shown by Damour et al[73] that a collagen-glycosaminoglycan-chitosan could allow a 60% take of CAE. This was demonstrated on one patient (Fig. 3.3).

Long-term evaluation has been done in a few reports. The aim of these studies was to analyze on skin biopsies the state of the dermal-epidermal junction,[50,51,74,75] the colonization of the graft by Langerhans cells, the revascularization and reconstruction of the neodermis. It is clear that the results are controversial. Several methodologies such as light microscopy, immunohistochemical analyses, morphometric approaches, electron microscopic and ultrastructural immunolabeling techniques have been performed de novo and show the presence of a complete basal lamina comprising mature hemidesmosomes within about 3 weeks. However, anchoring fibrils appear sparse and immature after 12 months. After 4-5 years, elastin expression is also observed in the CAE neodermis. These observations show that the CAE regenerates stable differentiated epidermis capable of inducing dermal regeneration.[74-77]

In contrast to these observations, Woodley et al[54,55] reported an abnormal quantity and quality of the anchoring complex elements 4 to 5 months after grafting. These drawbacks in the dermal-epidermal junction could be responsible for blisters.[46,55,78-80] These abnormalities may disappear with the restoration of correct dermal-epidermal junction within a few years.[50,67]

Long-term aesthetic results appeared to be slightly better with CAE than those observed with split-thickness skin graft. While skin appears flexible, smooth and pliable with fewer hypertrophic scars according to some authors,[35,54,55,62,78,81] it still shows hypertrophic scars for others, especially when CAE is placed on a deeply excised wound.[82] Moreover, it is now well established that long-term wound retraction can be very important, especially when dermis is lacking. This problem can lead to serious consequences in the case of children whose growth is hampered.[74]

We should also mention that within a single group, some selection criteria may have changed with time, i.e., increasing of the necessary minimal burn area. Results obtained over several years are means and do not reflect these changes. Indeed, the use of CAE requires intensive care and the participation of all clinical specialties. This technique has progressed permitting its use on much larger burn surfaces (Fig. 3.3). In the 1980s, it was frequently used for burn areas over 30%; and in 1996, it is recommended for burns over 60%. The percentage of mortality of CAE treated patients may possibly increase in the near future, not because of inadequacy of the sheets, but because of the severity of the disease. In spite of the progress made by different teams in terms of experience and ability, the mortality rate of extensive burn patients is not yet totally eradicated. However, it has tremendously decreased.

The advantage of this technique is that a small biopsy of skin can provide enough epidermal sheets to cover the entire body surface while avoiding the donor site wound. The initial enthusiasm generated by this technique[83] has been tempered by some authors because of its delay of obtention (≥3 weeks), its cost which increases with the number of regrafts, the demonstration of low initial engraftment rate, rehabilitation delays, graft fragility and delayed graft loss. Of all these

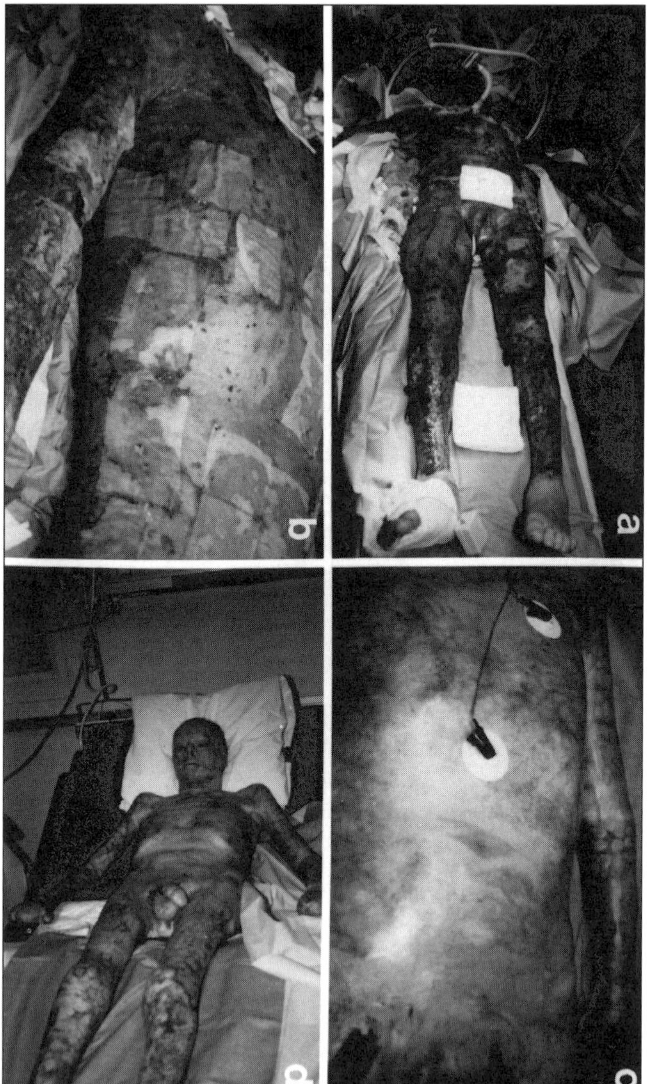

Fig. 3.3. Graft of cultured autologous epidermis (CAE) on a 96%, burned man. Only his feet were preserved by leather shoes. (a) Morphological aspect of the wound before grafting CAE. After excision to aponeurosis, his body was temporarily covered with homograft; observe remnants of dermis (white areas). (b) Morphological aspect of the covered site, 10 days post grafting (second take). (c) Morphological aspect 3 months later. (d) The patient 5 months after the burn; all the techniques were used for his coverage. His face was covered with secondly regenerated scrotum skin, thorax abdomen and thighs with CAE, legs and arms with a variant of combined techniques using small pieces of skin sampled from the toes, covered with CAE.

problems, the two most restrictive are delay of grafting, which is responsible for local infection, and prohibitive cost when they are industrially provided.[45,63,84-86] However, they are of great help to surgeons who need to cover extensively burned patients. CAE must be considered as an adequate complement for the treatment of deep extensively burned patients, but not as a miracle treatment on its own. Ten years of experience using CAE for the treatment of extensively burned patients has allowed us to propose a therapeutical strategy in which CAE can be used under several different modalities.

SURGICAL ASPECTS

Indication for the Use of Cultured Epidermis

It is clear that cultured epidermis has not replaced classical autografting. Its use is limited to only the most serious cases. Several conditions need to be met before considering its use.

Extent and Depth of Burns

Munster and Donati[38,41] consider patients as candidates for cultured epithelium if they present burns over 40% of the total body surface area (TBSA) and if they present a substantial third degree component. From our point of view, cultured autologous epithelia is recommended for cases presenting over 50% TBSA burns. The main difficulty remains in evaluating the depth of injury. Indeed, distinguishing a deep burn (third degree) from a superficial burn is still based on clinical evaluation. This method is totally subjective and errors are still frequent, especially in the case of back and limb roots.

Age and General State of Health

Clinical experience has shown that the use of cultured epidermis is possible with elderly patients only if they were in a good state of health prior to injury. The presence of preexisting cardiovascular or hepatic problems, which compromise the prognosis, is a definite contraindication. However, the presence or absence of inhalation injury is not a contraindication.[41,48]

The delay between burn injury and plating of the keratinocytes must be as short as possible. When this time exceeds 48 hours, bacterial contamination is inevitable and antibiotic treatment of the in vitro biopsy is required before processing. This further extends the day of grafting. We recommend that the biopsy sampling for cell culture be done as soon as the patient is admitted to the burn center.

Initial Surgical Treatment

The decision to use cultured epidermis involves a rigorous and precise therapeutic protocol. A strict chronology of events must be followed so that both cultured epidermis and optimal preparation of the wound bed are ready on the day of grafting.

The main steps in management are as follows: excision of all burned tissues, temporary coverage of excised areas, coverage of priority areas with autografts if possible, and maintenance of the general state of health with adequate nutrition and prompt treatment of infections.

Surgical treatment starts as soon as the patient is under general anesthesia. The patient is washed, brushed and totally shaved. In circumferential injuries to the extremities, there is a need to record pressure in the muscle compartment and escharotomy incisions are made when the pressure exceeds 40 mm Hg. Neck and chest burns are also subjected to escharotomies if thoracic expansion is restricted. The escharotomy incisions are covered with vaseline tulle plugs and stapled to the edges. The whole area is then covered with a thick layer of silver sulfadiazine (in the case of superficial burns). This is covered with gauze pads and bandaged. The dressing is changed every day at the patient's bed.

Surgical excision starts on the third day after the patient is stabilized. This delay allows time to precisely determine the burn depth. Early excision, within 24 hours, can be dangerous because of the hemodynamic instability of the patient and the difficulty in estimating burn depth.[38,87]

Excision of burned tissues must be performed in a surgical environment with regard to three guidelines: speed, limited loss of blood and ablation of all nonviable tissues. Due to surgical constraints, a maximum of 30% of the total body surface can be excised at one time.[41]

Promptness of treatment requires a trained team and participation of several operators. We practice excision of homogeneous areas in the following order: lower limbs, upper limbs, front chest and abdomen, back and buttocks. The face and hands are the last areas to be excised.

Excisions are performed every 2 or 3 days in order to completely remove the burned tissues within the first 10 days. Loss of blood is limited by the use of tourniquets on the limbs, by adrenaline and serum-soaked compresses and by elevation of the limbs.

From our point of view, there should be no controversy regarding the type of excision performed: excision to aponeurosis[1,2,88] or tangential excision. Both of these excision types have precise indications. Excision to aponeurosis is a rapid and simple surgical operation which must be performed only in distal segments of the limbs, in cases of deep burns, i.e., on forearms and legs. We also practice excision to aponeurosis in case of limited burns on elderly patients or those who are in a poor state of health (Fig 3.3). The main problem of excision to aponevrosis is the catastrophic aesthetic result. An argument against this technique is the disrespect for dermal tissue. In fact, preservation of the dermis in areas where its thickness allows it (back, limbs roots) will provide excellent aesthetic results and superior graft take.

We perform tangential excision with a Lagrot or Watson hand razor. The risk of tangential excision is that it tends to leave devitalized tissue which is a potential source of infection.

Temporary Coverage

Excised areas can be temporarily covered with several kinds of dermal substitutes. The older technique consists of using a greasy dressing with an antiseptic, which protects against dryness and infection but allows granulation tissue to develop. This kind of dressing is not well adapted for the use of cultured epidermis because it leads to excessive granulation tissue formation between the time of excision and cover with cultured epidermis.

An alternative method recommending the use of synthetic dermal substitutes such as Biobrane® and Comfeel®, which are set in place for several weeks, is generally well tolerated. These substitutes present an effective barrier against infection and granulation tissue formation.

Homografts from banks, according to Cuono[30,89] and our own experience, are the most reliable temporary substitutes. They are easy to use and effectively protect against infection and exudation. Cryoconserved homografts from skin banks are now accepted as indispensable temporary substitutes to cover the wound until the preparation of cultured epidermal grafts. We use graft of 0.2 to 0.3 mm depth from homogeneous preparations, taking adequate biosafety precautions. As more local banks are progressively being established the actual problems of low supplies will progressively be solved.

The homografts are expanded 1.5 fold and maintained on the wound surface under stretch with staples. They are then covered with greasy dressings and antibiotic ointment (neomycin). Homografts are regularly inspected and changed in the event of nontake or infection. Several thousand square centimeters are frequently used for temporary coverage and it is necessary to establish precise and dated diagrams of the different areas covered with homografts.

When homografts are not available, the use of extracellular matrices according to the Burke model[90] represents great hope and these matrices are now on the market.

PROTOCOL FOR THE USE OF CULTURED EPIDERMIS

Cutaneous Biopsy Sampling

The biopsy must be protected from bacterial contamination and the eventual effects of cytotoxic products. Sampling must be done in an aseptic environment used for surgical procedures.

For sampling, we choose areas that have rarely been exposed to ultraviolet light and have not been subjected to friction. The groin, axillary fold, perineum and scalp are frequently areas of choice for sampling. The area is then carefully washed with degreasing liquid, rinsed and disinfected. It is rinsed once again before sampling.

Depending on the procedure adopted by the cell culture laboratory, the biopsy will be either sampled with a dermatome (as for a split skin graft of 0.3 mm) or as a total skin sample with a hand bistoury. The sample is rapidly introduced into a special container for transportation and kept at 4°C until it can be processed in the cell culture laboratory.

Recipient Area for Cultured Epidermal Grafting

The most favorable areas are the front of the thorax, abdomen, and anterior, lateral and medial sides of the thighs and arms, representing approximately 30% of the TBSA for an adult.[35] These areas have priority for coverage with cultured epidermis as compared to posterior regions of the head, extremities and articulation zones.

Technique for Grafting Epidermal Sheets

Bacteriological studies of the wound bed are regularly performed and a biopsy is sampled a few days before grafting. In case of bacterial colonization, local treatment with an appropriate topical agent is instituted. Additional samples are investigated a day before the grafting to determine the local state. In the case of persistent local bacterial infection the epidermal grafting should be postponed for 48 hours.

On the day of epidermal grafting, areas which have been previously covered with homografts are superficially abraded with either a bistoury or an abrasive sponge. We strongly recommend tangential excision with a hand razor. In this manner, we hope to preserve a wound bed containing some dermal constituents. Efficient hemostasis can be obtained by applying adrenaline-soaked compresses. After surgical preparation, the operative field is totally redraped, the asepsis conditions are repeated and the surgical staff aseptically dressed. The priority areas to be grafted are evaluated in comparison to the total amount of tissue available. The use of patchworks of cultured epidermis in a scattered manner must be avoided. The sheets are removed from their adapted transportation container and applied on a smooth and vascularized, but nonhemorrhagic and noninfected, area. No secondary displacement is allowed. Sheets are located conjointly without empty spaces in between and fixed with metallic staples at each corner of the protection gauze. A large piece of dry tulle is stapled over the grafted area, under a stretch, to improve retention. The entire area is then covered with thick gauze compresses and bandaged.

Post-Surgical Surveillance

Whereas the gauze compresses are changed every 48 hours, the transfer gauze and the dry tulle used for retention are not changed for one week. After 8 days, the dry tulle and transfer gauzes are removed if they detach easily. Exudation and thick outflow of pus during the dressing are signs of bacterial contamination which compromises the graft take. It is necessary to perform bacteriological samples because an infection at this stage with either *Staphylococus aureus* or *Pseudomonas aeruginosa* will induce irreversible destruction of the graft. With clinical evidence of infection the grafted area is treated with suitable antimicrobial soaked compresses. Antimicrobial dilutions must be proper because of their potential cytotoxicity, i.e., colimycin 100 mg/L polymixin 1 million units/L.

The superficial aspect of the grafted epidermis after removal of the transfer gauzes resembles a glazed membrane, similar to a nonepidermized wound. Care must be taken to prevent the epidermis from pulling off. Gradually, the cultured epidermis will assume the dull look of a thin dermal-epidermal graft. Desquamation of the superficial layer, which always occurs between the 15th and 20th day, does not compromise the strength of the epidermis. Even a partial take of epidermal grafts can lead to complete wound healing by progressive epidermization from the islets. As we already mentioned, the extreme fragility of the cultured epidermis within the first few weeks must be absolutely considered and any rough manipulations must be avoided.

Long-term observations show weak inflammatory reactions and rare development of hypertrophic scarring of the grafts. However, the grafted area has a tendency to retract, causing serious problems in the case of children for whose growth is hampered. In this particular situation, we apply a "combined technique" which is different from the sandwich technique.

COMBINED TECHNIQUE

Although allografts remain the best skin substitute and the best preparation of the wound bed before grafting with cultured epithelium, we had to reduce the use of allografts because in France, the availability of allografts is an acute problem, which greatly limits their use. Secondly, the infectious risk related to xenogeneic or allogeneic products must be kept in mind. That is why we progressively developed a "combined technique" which associates a widely expanded autograft (1 x 6) and cultured keratinocytes sheets overlay. Only a few reports previously described this grafting procedure.[35]

When cultured keratinocytes are ready to graft (2 to 3 weeks after burns), donor sites are usually ready for a second harvesting. The combined technique uses these two means of coverage together. After a careful cleaning of the excised wound bed, polyvidone iodine impregnated gauzes are applied for 10 minutes. The wound bed is then rinsed with serum and dried. Widely expanded autografts (1 x 6) are then applied and covered with the epithelial sheets. Then, a dry dressing is used to protect the implant. The graft take is evaluated on day seven. In order to prevent or cure local infection, appropriate antiseptics and antibiotics are often used.[91,92]

We have applied this technique especially in children (Fig 3.4). Since 1990, 12 children have undergone this combined method. None died. The average age was 10 years, ranging from 2 to 15 years. The average burned surface was 65% of the body surface (range 42 to 88).

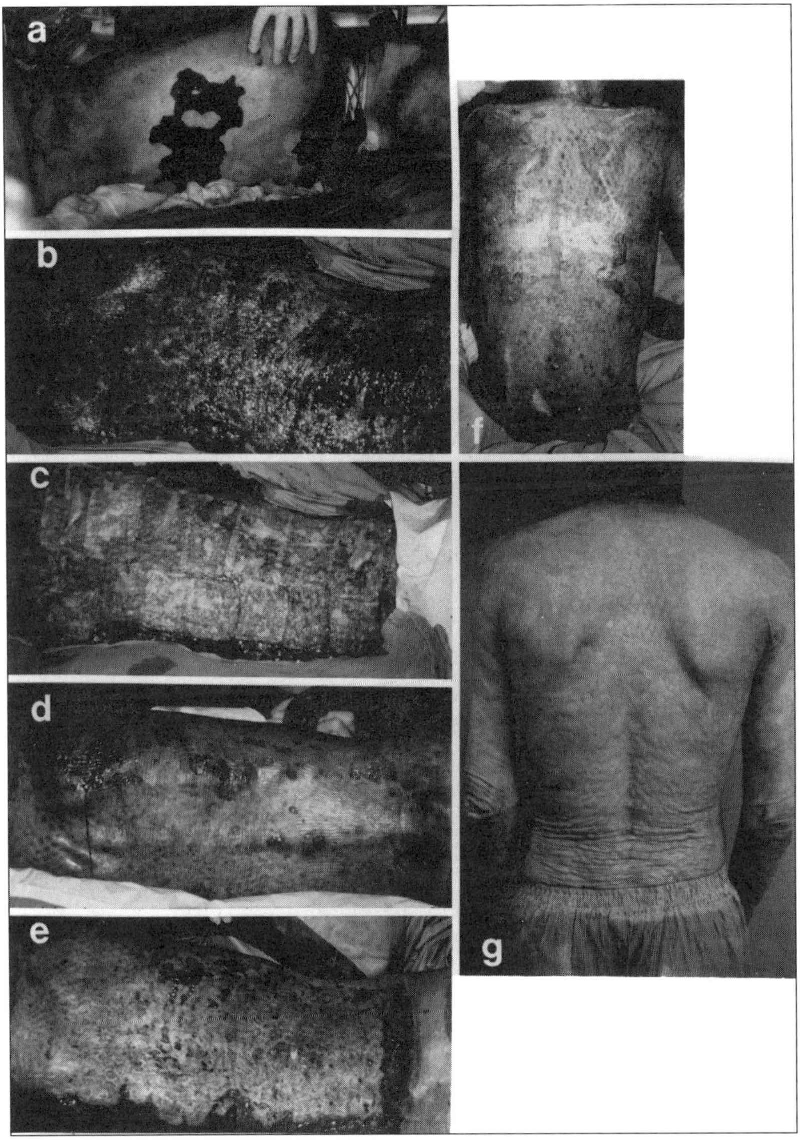

Fig. 3.4. Graft of cultured autologous epidermis (CAE) on a large mesh autograft according to combined technique on a 14 year old boy. (a) Morphological aspect of the wound immediately after the disaster. (b) Morphological aspect of the wound just before the graft. (c) Graft CAE on a 1 x 6 meshed autograft. (d) Morphological aspect of the wound 3 weeks after grafting. (e) Morphological aspect of the wound site 7 weeks after grafting. (f) Morphological aspect of the same site 3 months after grafting. (g) Morphological aspect of the covered site,1 year after grafting.

The average deep burn surface was 60% (range 42 to 88). This technique alone led to the complete healing of 30% of the total body surface on average, with an initial take rate of 84%. The association of the CEA and widely meshed autograft resulted, on the take off day, in a continuous epidermalization. No secondary loss was noticed. In seven cases, additional late stamp grafts were applied to complete the coverage.

The average length of the stay in the reeducation center was 10.5 months. All the children returned to a normal family and social life. Among them, ten have been clinically evaluated with an average delay of three years. The children who were of school age at the time of the burn are now one year behind at school. The areas which received the combined technique seem to present less retraction than areas grafted with cultured keratinocytes alone (Fig. 3.4).

Widely meshed autografts often fail because the epidermis is slow to grow in the holes of the network. This delay allows the development of local infection and granulation tissue. The vulnerability of large meshed autograft has led to development of the "sandwich" method:[93] a wide mesh autograft is covered with 1.5:1 meshed allograft. The allograft protects the autograft against physical and infectious aggression while it develops. When cultured sheets are applied on 6:1 autograft, the epidermization is complete at the removal of the gauze on day seven. Thanks to the cultured epithelia, the granulation tissue is not allowed to develop in the mesh of the widely expanded autograft. The epithelium seems more resistant to infection or mechanical trauma than when only the combined technique is used: we observed no secondary loss of the grafts. It appears that a synergy may occur between the two means of grafting.

The meshed autograft, even when widely expanded, provides a thin dermal network which may play a key role in cellular interactions between fibroblasts and keratinocytes. The importance of replacing the dermis in third degree burns is commonly admitted for both the immediate results and the restoration of skin elasticity.

The trophic role of cultured keratinocytes by secreting growth factors and cytokines is well known. The association of widely expanded autograft and allograft cultured keratinocytes also seems promising.[94]

The combined technique associating widely meshed expanded autograft and cultured keratinocytes provides secure and resistant coverage in a single operative procedure.

OTHER APPLICATIONS

In plastic surgery, cultured keratinocytes have been used to cover large skin defects after surgical excisions, for giant congenital naevi,[95] to cover the cavity after mastoidectomy,[96] to graft the separation site in combined twins,[97] for tattoo removal,[98] degloving injury[99] and for the treatment of hypospadias.[100]

Allogeneic keratinocytes have been reported to accelerate the treatment of extensive second degree burns[57,101] and of graft donor sites.[58] These allogeneic cells are later on replaced by recipient cells.[102]

Autologous or allogeneic cultured epithelia are also used to stimulate the healing of chronic wounds.[103-108] The place of cultured epithelia for the treatment of junctional epidermolysis bullosa was described by Carter,[109] Schofield[110] and McGrath.[111] They may be promising indications for oral resurfacing in maxillofacial surgery.[112,113]

CONCLUSION

The treatment of extensively burned patients requires a close cooperation between health care teams with an access to advanced techniques. Cultured epidermis represents one of these technical tools and it would be wrong to believe that this technical progress leads to a simplification of the management of a seriously burned patient.

The use of cultured epidermis requires a rigorous therapeutic strategy which requires the participation of various personnel including cell culture biologists, who become essential partners. In addition to the technical aspect, the new therapeutic possibility of cultured epidermis has allowed us to be hopeful when faced with the most severe cases.

Keratinocyte culture provides a barrier function that assures definitive coverage. However, it does not permit total regeneration of the skin "organ" but allows for the formation of scar tissue. If CAE helps the patient to survive, dermis plays an important functional and aesthetic role. Since the dermis is now commercially available in US and soon in Europe, indications and instructions for its use need to be developed. The ideal perspectives of skin reparation are the dermal-epidermal models.

References

1. Jackson P, Loughrey CM, Lightbody JH et al. Effect of hemodialysis on total antioxidant capacity and serum antioxidants in patients with chronic renal failure. Clin Chem 1995; 41:1135-1138.

2. Janzekovic Z. A new concept in the early excision and immediate grafting of burns. J Trauma 1970; 10:1103-1108.
3. Brown JB, Fryer MR, Randall P et al. Postmortem homografts as biological dressings for extensive burns and denuded areas. Ann Surg 1953; 138:618-630.
4. Bromberg BE, Song C, Mohn MP. The use of pigskin as a temporary biological dressing. Plast Reconstr Surg 1965; 36:80-90.
5. Prunieras M. L'épiderme et les cultures de cellules épidermiques. Editor SPEI, Département Médical 1965; 1-141.
6. Basset C, Evans V, Earle W. Characteristics and potentials of long term culture of human skin. Plast Reconst Surg 1956; 17:421-429.
7. Earle W, Brian J, Schellinge E. Certain factor limiting the size of the tissue culture and the development of massive culture. Ann N Y Acad Sci 1954; 58:1000-1011.
8. Puck T. Clonal growth of mamalian cells in vitro: growth characteristics of colonies from single HeLa cells with a "Feeder Layer". J Exp Med 1956; 103: 273-283.
9. Karasek M. Dermal factor affecting epidermal cells in vitro. J Invest Dermatol 1972; 59:99-101.
10. Rheinwald JG, Green H. Serial cultivation of strains of human epidermal keratinocytes: the formation of keratinizing colonies from single cells. Cell 1975; 6:331-344.
11. Green H, Kehinde O, Thomas J. Growth of cultured human epidermal cells into multiple epithelia suitable for grafting. Proc Natl Acad Sci USA 1979; 76:5665-5668.
12. O'Connor NE, Mulliken JB, Banks-Schlegel S et al. Grafting of burns with cultured epithelium prepared from autologous epidermal cells. Lancet 1981; I: 75-78.
13. Gallico GG, O'Connor NE, Compton CC et al. Permanent coverage of large burn wounds with autologous cultured human epithelium. N Engl J Med 1984; 311:448-451.
14. Ham RG. Clonal growth of mammalian cells in a chemically defined, synthetic medium. Microbiology 1965; 53:288-293.
15. Gilchrest BA, Calhoun JK, Maciag T. Attachement and growth of human keratinocytes in a serum-free environment. J Cell Physiol 1982; 112:197-206.
16. Boyce ST, Ham RC. Calcium-regulated differentiation of normal epidermal keratinocytes in chemically defined clonal culture and serum-free serial culture. J Invest Dermatol 1983; 81:33S-40S.
17. Pittelkow MR, Scott RE. New techniques for the in vitro culture of human skin keratinocytes and perspectives on their use for grafting of patients with extensive burns. Mayo Clin Proc 1986; 61:771-777.
18. Daniels JT, Kearney JN, Ingham E. Human keratinocyte isolation and cell culture: a survey of current practices in the UK. Burns 1996; 22(1):35-39.

19. Blacker KL, Williams ML, Goldyne M. Mitomycin C-treated 3T3 fibroblasts used as feeder layers for human keratinocyte culture retain the capacity to generate cicosanoids. J Invest Dermatol 1987; 89:536-540.
20. Green H, Rheinwald JG, Sun TT. Properties of an epithelial cell type in culture: the epidermal keratinocyte and its dependence on products of the fibroblast. Prog Clin Biol Res 1977; 17:493-500.
21. Barrandon Y, Green H. Cell size as a determinant of the clone-forming ability of human keratinocytes. Proc Natl Acad Sci USA 1985; 82:5390-5394.
22. Limat A, Hunziker T, Boillat C et al. Post mitotic human dermal fibroblasts efficiently support the growth of human follicular keratinocytes. J Invest Dermatol 1989; 92:758-762.
23. Barrandon Y, Green H. Three clonal types of keratinocytes with different capacities for multiplication. Proc Natl Acad Sci USA 1987a; 84:2302-2306.
24. Barrandon Y, Green H. Cell migration is essential for sustained growth of keratinocyte colonies: the roles of transforming growth factor-α and epidermal growth factor. Cell 1987b; 50:1131-1137.
25. Rheinwald JG, Green H. Epidermal growth factor and the multiplication of cultured human epidermal keratinocytes. Nature 1977; 265:421-424.
26. Green H. Cyclic AMP in relation to proliferation of the epidermal cell: a new view. Cell 1978; 15: 801-815.
27. Adams JC, Watt FM. Changes in keratinocyte adhesion during terminal differentiation: reduction of fibronectin binding precedes $\alpha 5\beta 1$ integrin loss from the cell surface. Cell 1990; 63:425-435.
28. Green H, Kehinde O. Transplantable sheet of living keratinous tissue. Patent 4, 304,866,dec 1981.
29. Kitano J, Okada N. Separation of the epidermal sheets by dispase. Br J Dermatol 1983; 108:555-558.
30. Andreassi L, Casini L, Trabucchi E et al. Human keratinocytes cultured on membranes composed of benzyl ester of hyaluronic acid suitable for grafting. Wounds 1991; 3(3):16-126.
31. Ronfard V, Broly H, Mitchell V et al. Use of human keratinocytes cultured on fibrin glue in the treatment of burn wounds. Burns 1991; 17(3):181-184.
32. Kaiser HW, Stark GB, Kopp J et al. Cultured autologous keratinocytes in fibrin glue suspension, exclusively and combined with STS-allograft (preliminary clinical and histological report of a new technique). Burns 1994; 20(1):23-29.
33. Eldad A, Burt A, Clarke JA. Cultured epithelium as a skin substitute. Burns 1987; 13:173-180.
34. De Luca M, Albanese E, Bondanza S et al. Multicentre experience in the treatment of burns with autologous and allogeneic cultured epithelium, fresh or preserved in a frozen state. Burns 1989; 15(5): 303-309.

35. Teepe RGC, Kreis RW, Koebrugge EJ et al. The use of cultured autologous epidermis in the treatment of extensive burn wounds. J Trauma 1990; 30:269-275.
36. Foyatier JL, Faure M, Hezez G et al. Application clinique des greffes d'épiderme cultivé chez le brûlé. Ann Chir Plast Esthet 1990; 35(1):39-46.
37. Tamisani AM, Ferretti S, Sangiorgio L. Critical reflections on the use of human cultured keratinocytes in children with burns. Eur J Pediatr Surg 1992; 2:223-226.
38. Donati L, Magliacani G, Bormioli M et al. Clinical experiences with keratinocyte grafts. Burns 1992; 18:19S-25S.
39. Siwy BK, Compton CC. Cultured epidermis: Indiana University Medical Center's Experience. J Burn Care Rehabil 1992; 13:130-137.
40. Weekley R, Klein R. Clinical nursing experience with cultured epidermal autografts. J Burn Care Rehabil 1992; 13:138-141.
41. Munster AM, Weiner SH, Spence RJ. Cultured epidermis for the coverage of massive burn wounds. Ann Surg 1990; 211:676-680.
42. Haith LR, Patton ML, Goldman WT. Cultured epidermal autograft and the treatment of the massive burn injury. J Burn Care Rehabil 1992;13:142-146.
43. Nave M. Wound bed preparation: approaches to replacement of dermis. J Burn Care Rehabil 1992; 13:147-153.
44. Barillo DJ, Nangle ME, Farrell K. Preliminary experience with cultured epidermal autograft in a Community Hospital Burn Unit. J Burn Care Rehabil 1992; 13:158-165.
45. Herndon DN, Rutan LR. Comparison of cultured epidermal autograft and massive excision with serial autografting plus homograft overlay. J Burn Care Rehabil 1992; 13:154-157.
46. McAree K, Klein RL, Boeckman CR. The use of cultured epithelial autografts in the wound care of severely burned patients. J Pediatr Surg 1993; 28(2):166-168.
47. Still JM, Orlet HK, Law EJ. Use of cultured epidermal autografts in the treatment of large burns. Burns 1994; 20:539-541.
48. Latarjet J, Grangolphe M, Hezez G et al. The grafting of burns with cultured epidermis as autografts in man. Scand J Plast Reconstr Surg Hand Surg 1987; 21:241-244.
49. Faure M, Mauduit C, Schmitt D et al. Growth and differentiation of human epidermal cultures used as auto- and allografts in humans. Br J Dermatol 1987; 116:161-170.
50. Compton CC, Gill JM, Bradford DA et al. Skin regenerated from cultured epithelial autografts on full-thickness burn wounds from 6 days to 5 years after grafting. Lab Invest 1989; 60(5):600-612.
51. Neveux Y, Rives JM, Le Breton C et al. Clinical interest of cutaneous models reproduced in vitro for severe burn treatment: histopathological and ultra-structural study. Cell Biol Toxicol 1995; 11: 173-178.

52. Deglise B, Benathan M, Frenk E et al. Résultats préliminaires du traitement des brûlés par autogreffe d'épidermes de culture. Schweiz Med Wochenschr 1987; 117:1380-1393.
53. Heimbach DM. A nonuser's questions about cultured epidermal autograft. J Burn Care Rehabil 1992; 13:127-129.
54. Woodley DT. Covering wounds with cultured keratinocytes. JAMA 1989; 262:2140-2141.
55. Woodley DT, Peterson HD, Herzog SR et al. Burn wounds resurfaced by cultured epidermal autografts show abnormal reconstitution of anchoring fibrils. JAMA 1988; 259:2566-2571.
56. Desai MH, Mlakar JM, McCauley RL. Lack of the long term durability of cultured keratinocytes burn-wound coverage. J Burn Care Rehabil 1991; 12:540-545.
57. De Luca M, Bondanza S, Cancedda R et al. Permanent coverage of full skin thickness burns with autologous cultured epidermis and re-epithelialization of partial skin thickness lesions induced by allogeneic cultured epidermis: a multicentre study in the treatment of children. Burns 1992; 18:16S-18S.
58. Myers S, Navsaria H, Sanders R et al. Transplantation of keratinocytes in the treatment of wounds. Am J Surg 1995; 170:75-83.
58. Sheridan RL, Tompkins RG. Cultured autologous epithelium in patients with burns of ninety percent or more of the body surface. J Trauma 1993; 38(1):48-49.
59. Odessey R. Addendum: Multicenter experience with cultured epidermal autograft for treatment of burns. J Burn Care Rehabil 1992; 13:174-180
60. Braye F, Oddou L, Bertin-Maghit M et al. Use of widely meshed autograft and cultured epithelial autografts in the treatment of major burns in children. Br J Surg (submitted).
61. Still J, Scheirer R, Law E. Acute cholecystectomie performed through cultured epithelial autografts in patient with burn injuries: a case report. J Burn Care Rehabil 1996; 17:429-431.
62. Krupp S, Benathan M, Meult M et al. Current concepts in pediatric burn care: management of burn wounds with cultured epidermal autografts. Eur J Pediatr Surg 1992; 2:210-213.
63. Rue LW III, Cioffi WG, McManus WF et al. Wound closure and outcome in extensively burned patient treated with cultured autologous keratinocytes. J Trauma 1993; 34:662-668.
64. Burke JF, Quinby WC, Bondoc CC. Primary excision and prompt grafting as a routine therapy for the treatment of thermal burns in children. Surg Clin North Am 1976; 56:477-494.
65. Zawachi BE. Temporary wound closure after burn excision. J Burn Care Rehabil 1986; 7:138-143.
66. Leary T, Lloyd P, Appleby M et al. Epidermal keratinocyte self-renewal is dependent upon dermal integrity. J Invest Dermatol 1991; 99:422-430.

67. Mommaas M, Teepe RG, Leigh IM et al. Ontogenis of the basement membrane zone after grafting cultured human epithelium: a morphogenic and immunoelectron microscopic study. J Invest Dermatol 1992; 99:71-77.
68. Barr M, Linwood FT. Various sugars and polyols on the growth of micro-organisms. J Am Pharm Ass 1957; 46:219-221.
69. Martin B, Linwood FT. A study of the inhibitory concentration of glycerine-sorbitol and propylene-sorbitol combination on the growth of micro-organisms. J Am Pharm Ass 1957; 46:217-218.
70. Basile ARD. A comparative study of glycerinized and lyophilized porcine skin dressings for third degree burns. Plast Reconstr Surg 1982; 69:969-974.
71. Yannas IV, Burke JF. Design of an artificial skin: I. Basic design principles. J Biomed Mat Res 1980a; 14:65-81.
72. Yannas IV, Hansbrough JF, Ehrlich HP. What criteria should be used for designing artificial skin replacements and how well do the current grafting materials meet these criteria? J Trauma 1984; 24:29S-39S.
73. Damour O, Fabreguette A, Augustin C et al. D'une meilleure compréhension des mécanismes de la cicatrisation à l'utilisation des substituts cutanés dans le traitement des plaies difficiles. J Plaies Cicatrisation 1996; 7-12.
74. Compton CC. Current concepts in pediatric burn care: the biology of cultured epithelial autografts: an eight-year study in pediatric burn patients. Eur J Pediatr Surg 1992; 2: 216-222.
75. Franzi AT, D'Anna F, Zicca A et al. Histological evaluation of human cultured epithelium before and after grafting. Burns 1992; 1:26S-31S.
76. Cairns BA, De Serres S, Peterson HD et al. Skin replacements. The biotechnological quest for optimal wound closure. Arch Surg 1993; 128:1246-1252.
76. Nanchahal J, Ward CM. New grafts for old? A review of alternatives to autologous skin. Br J Plast Surg 1992; 45:354-363.
77. Teepe RGC, Burger A, Ponec M. Immunohistochemical studies on regeneration in cultured epidermal autografts used to treat full-thickness burn wounds. Clin Exp Dermatol 1994; 19:16-22.
78. Kumagai N, Nishina H, Tanabe H et al. Clinical application of autologous cultured epithelia for the traitment of burn wounds and burn scars. Plast Reconstr Surg 1988; 82(1):99-110.
79. Petersen MJ, Lessane B, Woodley DT. Characterization of cellular elements in healed cultured keratinocyte autografts used to cover burn wounds. Arch Dermatol 1990; 126:175-180.
80. Merrick P, Meyer AA, Herzog S et al. Scanning electron microscopy of cultured human keratinocytes. J Burn Care Rehabil 1990; 11: 228-236.
81. O'Connor NF, Gallico GG, Kehinde O et al. Grafting of burns with cultured epithelium prepared from autologous epidermal cells: II.

Intermediate term results on three pediatric patients. In: Hunt TK, Heppenstall HBS, Pines E et al, eds. Soft and Hard Tissue Repair: Biological and Clinical Aspects. New York: Praeger Publishers 1994:283-292.
82. Warden G. The fifth quinquennium: 1989-1993. J Burn Care Rehabil 1993; 14:247-251.
83. Fisher JC. Skin the ultimate solution for the burn wound. N Engl J Med 1984; 311: 466-467.
84. Tompkins RG, Hilton JF, Burke JF et al. Increased survival after massive thermal injuries in adults: preliminary report using artificial skin. Crit Care Med 1989; 17(8):734-740.
85. Sheridan RL, Hegarty M, Tompkins RG et al. Artificial skin in massive burns—results to ten years. Eur J Plast Surg 1994; 17:91-93.
86. Germain L, Auger FA. Tissue engineered biomaterials: biological and mechanical characteristics. In: Wise D, Trantolo DJ, Altobelli DE, Yaszemski MJ, Gresser JD, Schwartz ER, eds. Encyclopedic Handbook of Biomaterials and Bioengineering, Part B; Applications. New York, USA: Marcel Dekker, Inc. Publishers, 1995; 699-734.
87. Teepe RGC, Ponec M, Kreis RW et al. Improved grafting method for treatment of burns with autologous cultured human epithelium (letter). Lancet 1986; I:385.
88. Jackson DM. Excision and closure of the wound as applied to burns. Proc R Soc Med 1972; 65:23-24.
89. Cuono C. Use of cultured epidermal autografts and dermal allografts as skin replacement after burn injury. Lancet 1986; I:75-78.
90. Burke JF, Yannas IV, Quinby WC et al. Successful use of a physiologically acceptable artificial skin in the treatment of extensive burn injury. Ann Surg 1981; 194:413-428.
91. Cooper ML, Boyce ST, Hansbrough JF. Cytotoxicity to cultured human keratinocytes of topical antimicobial agents. J Surg Res 1990; 48:190-195.
92. Damour O, Zhi Hua S, Lasne F et al. Cytotoxicity evaluation of antiseptics and antibiotics on cultured human fibroblasts and keratinocytes. Burns 1992; 18(6):479-485.
93. Alexander JM, McMillian BG, Law E et al. Treatment of severe burns with widely meshed skin autograft and meshed ski allograft overlay. J Trauma 1981; 21:433-438.
94. Duinslaeger L. Short and long term effect of allogenic keratinocytes and derivates in the treatment of burn wounds. State of the Art Control of Burn Scar Symposium of EBA Amsterdam, Oct 1996.
95. Gallico G, O'Connor NE, Compton C et al. Cultured epithelial autografts for giant congenital naevi. Plast Rcconstr Surg 1989; 84:1-9.
96. Premachandra DJ, Woodward BM, Milton CM et al. Treatment of post operative otorrhoea by grafting of mastoid cavities with cultured autologous epidermal cells. Lancet 1990; i:365-367.

97. Higgins CR, Navsaria HA, Stringer M et al. Use of two stage keratinocyte- dermal grafting to treat the separation site in conjoined twins. J R Soc Med 1994; 87:108-109.
98. Abbes M, Bourgeon Y, Regnier M et al. Notre expérience dans l'utilisation des cultures de cellules épidermiques en chirurgie réparatrice. Ann Chir Plast Esthet 1990; 35(1):31-38.
99. Clarke JA. Cultured skin for burn injury. Lancet 1986; 2:809.
100. Romagnoli G, De Luca M, Faranda F et al. Treatment of posterior hypospadias by the autologous graft of cultured urethral epithelium. N Engl J Med 1990; 323:527-530.
101. Phillips TJ, Gilchrest BA. Clinical applications of cultured epithelium. Epith Cell Biol 1992; 1:39-46.
102. Van Der Merwe AE, Mattheyse FJ, Bedford M et al. Allografted keratinocytes used to accelerate the treatment of burn wounds are replaced by recipient cells. Burns 1990; 16(3):193-197
103. Hefton JM, Weksler M, Parris A et al. Grafting of skin ulcers with cultured autologous epidermal cells. J Am Acad Dermatol 1986; 14:399-405.
104. Leigh IM, Purkis PE. Culture grafted leg ulcers. Clin Exp Dermatol 1986; 11:650-652.
105. Hunyadi J, Farkas B, Bertényi C et al. Keratinocyte grafting: a new means of transplantation for full-thickness wounds. J Dermatol Surg Oncol 1988; 14(1):75-78.
106. Phillips TJ, Gehinde O, Green H et al. Treatment of skin ulcers with cultured epidermal allografts. J Am Acad Dermatol 1989a; 21:191-199.
107. Phillips TJ, Gilchrest BA. Cultured allogeneic keratinocyte grafts in the management of wound healing. Dermatol Surg Oncol 1989b; 15:1169-1176.
108. Phillips TJ, Bhawan J, Leigh IM et al. Cultured epidermal autografts and allografts: a study of differentiation and allograft survival. J Am Acad Dermatol 1990; 23:189-198.
109. Carter DM, Lin AN, Varghese MC et al. Treatment of junctional epidermolysis bullosa with epidermal autografts. J Am Acad Dermatol 1987; 17:246-250.
110. Schofield OMV, Casella JP, Navsaria HA et al. Cultured keratinocyte allografts in dystrophic epidermolysis bullosa: preliminary observations. Br J Dermatol 1990; 123:66A.
111. McGrath JA, Schofield MV, Yamamoto AI et al. Cultured keratinocyte allografts and wound healing in severe recessive dystrophic epidermolysis bullosa. J Am Acad Dermatol 1993; 29(3):407-419.
112. Langdon J, Williams DM, Navsaria HA et al. Autologous keratinocytes grafting: a new technique for intra oral reconstruction. Br Dent J 1991; 171:87-90.
113. Ueda M, Ebata K, Kaneda T. In vitro fabrication of bioartificial mucosa for the reconstruction of oral mucosa. Basic research and clinical applications. Ann Plast Surg 1991; 27:540-549.

CHAPTER 4

Cultured Skin for Wound Closure

Steven T. Boyce

INTRODUCTION

Medical Needs

Skin substitutes for wound closure serve patients who are not candidates for surgical procedures with skin autograft. These patients include victims of full-thickness burns involving a very large total body surface area (TBSA), or patients with healing impairment who cannot donate skin for grafting. Generally, burns represent acute wounds, and healing impairment represents chronic wounds. However, these diseases may derive from etiologies including trauma, vascular disease, diabetes, or loss of ambulation. Estimates for hospitalizations from burns range from 60,000-80,000, and costs of recovery from acute injuries range from $36,000-117,000 per patient.[1] Pressure ulcers in the United States are estimated to have a prevalence of 9.2% in hospitalized patients[2] and to incur an annual cost of treatment exceeding $1.3 billion.[3] In addition, considerable morbidity results both from scarring[4] after grafting of excised burns and from pain, infection and loss of mobility in chronic wounds. Therefore, the medical and economic needs are very great for improved outcome from treatment of acute and chronic wounds.

Historic Standards of Wound Treatment

Split-thickness skin autograft remains the prevailing standard for treatment of excised, full-thickness burns. Typical expansion of

Skin Substitute Production by Tissue Engineering: Clinical and Fundamental Applications, edited by Mahmoud Rouabhia.

autograft involves meshing[5] at a ratio of 1:4, although ratios up to 1:9 have been reported with overlay of a biologic dressing such as cadaveric allograft.[6] Because the interstices of the mesh heal by secondary outgrowth from the edges of the meshed skin, scarring is directly proportional to mesh ratio, with minimal scarring observed in sheet grafts.[7] Consequently, sheet grafts are recommended for burns to the face and preferred for grafting of the hands.[8] Donor sites for split-thickness skin graft can be harvested usually 4-5 times over 2 months or longer. Therefore, the total expansion ratio of donor site for coverage of burn area may be estimated to be approximately 1:20. In cases of full-thickness burns over 60% or greater of the TBSA, wound closure is delayed by insufficient availability of donor skin for autografting, and long-term morbidity often results from scarring of the grafted wounds and also the donor sites. Greater availability of efficacious skin grafts would prospectively reduce morbidity from scarring. Additional skin grafts are required for reconstructive surgery after burns in which donor site morbidity is considered less problematic than the loss of mobility from scar contracture. In treatment of burn scars by reconstructive surgery, skin donation may also be performed by use of tissue expanders,[9,10] or skin flaps.[11] Morbidity from these sources of skin donation is minimal because grafts are full-thickness, and are closed by adjoining and suturing of the wound edges.

Treatment of chronic wounds may involve skin grafting, but availability of skin grafts is not usually limiting. More commonly in chronic wounds, the general health of the patient restricts surgical intervention, or the probability of healing of the donor site may be low. Conventional therapy requires debridement and cleansing of the wound, infection control and frequent changes with appropriate dressings. Many chronic wounds respond to this therapy and heal completely. Unna boots protect and apply compression to ulcers of the lower extremity to assist circulation and stimulate healing.[12] Subsequent management of the affected site, and of the general health (nutritional, psycho-social) condition often results in permanent wound closure. However, if conventional therapy fails, surgical intervention with split-thickness skin grafts is next in the algorithm of interventions.[13] Frequency of graft failure in treatment of chronic wounds is higher than in treatment of acute wounds, due to coincident deficiencies of vascular perfusion of the chronic wound. Wounds in which skin autograft fails may require vascular repair before healing can proceed.[14,15] Failure to close the chronic wound is rarely life-threatening, but may contribute to loss of mobility and dependence on others for performance of personal care.

Limitations of Conventional Treatments

For burns, the primary limitation to effective wound closure is availability, with a secondary consideration of donor site morbidity from scarring. In addition, donor sites can become infected (i.e., *Pseudomonas aeruginosa*) and convert from partial thickness wounds into full-thickness wounds. Availability of unlimited quantities of skin for grafting would prospectively minimize scarring during wound healing, and obviate complications from donor sites. In cases of large burns, increased availability may also reduce numbers of surgical procedures and hospitalization time.

For chronic wounds, it may be expected that wounds which fail to heal with conventional skin grafts will also fail to heal with alternative materials for skin repair. However, in wounds that respond to debridement and dressing changes, the time of healing may be reduced by the application of alternative materials, especially by the replacement of an epithelial cover. In fact, cultured allogeneic keratinocytes have been shown to promote closure of chronic wounds.[16-20] Therefore, reduction in time for wound closure and associated morbidity are the principal goals for advanced care of chronic wounds.

Current Alternatives

In response to the remaining limitations in care and closure of full-thickness skin wounds, a wide variety of materials has been developed and tested for skin repair. For burns, it is now recognized by consensus that both the epidermis and the dermis are required for an outcome with acceptable function and cosmesis.[21] Similarly, in chronic wounds, both the epidermis and the viable fibrovascular tissue must be restored to recover function. A summary of materials is presented in Table 4.1. Epidermal substitutes most often consist of cultured keratinocytes of autologous, allogeneic or chimeric populations.[22-24] Because culture proceeds by an exponential increase of cell numbers (2^n, where n = population doublings), an approximate 1×10^3 increase occurs in 10 days, and 1×10^6 occurs in 20 days. Therefore, skin

Table 4.1. Tissue substitutes for skin repair

Dermal	Epidermal
Acellular cadaveric matrix	Autologous cultured keratinocytes
Collagen-GAG ± fibroblasts	Allogeneic cultured keratinocytes
Collagen gel + fibroblasts	Thin epidermal graft
PLA/PGA	Epidermal suction blisters
PEO/PBT	Epidermal cell suspension

Legend: GAG, glycosaminoglycan; PLA, poly-lactic acid; PGA, poly-glycolic acid; PEO, polyethylene oxide; PBT, polybutylene terephthalate

substitutes containing cultured keratinocytes offer a categorical advantage in availability over other techniques for expansion of tissue. Other epidermal substitutes include: suction blisters;[25] cell suspensions;[26] and thin autografts.[27,28] Dermal substitutes include: cadaveric allodermis, either viable[29,30] or acellular;[31,32] lyophilized collagen-based sponges, without[28,33] or with[34,35] cultured fibroblasts; hydrated collagen gels with fibroblasts; or, synthetic biopolymers such as polylactic/polyglycolic acids (PLA/PGA)[36,37] and polyethylene oxide/polybutylene terephthalate (PEO/PBT).[38,39] These materials have been applied in a variety of combinations. De-epidermized cadaveric allograft has served as a dermal substitute in combination with epidermal suction blisters,[25] or autologous cultured keratinocytes.[29,30] Cultured allogeneic keratinocytes have been shown not to close full-thickness burns,[40,41] but have been reported to reduce scarring and facilitate healing of widely meshed autograft[42] and donor sites.[43] Composite grafts of collagen gels populated with allogeneic keratinocytes and fibroblasts have been reported to facilitate closure of chronic wounds,[44] although the allogeneic keratinocytes do not persist. Based on these favorable results, it is postulated that allogeneic keratinocytes release cytokines and extracellular matrix into the wound that stimulate healing by fibrovascular and epithelial cells at the site.[45,46] It has also been demonstrated that culture of allogeneic keratinocytes depletes resident leukocytes in donor tissue,[47,48] and reduces expression of class II histocompatibility antigens (HLA-DR) on keratinocytes.[49-51] Therefore, it is believed that immunogenicity of cultured epithelium is reduced compared to allogeneic epidermis of cadaver skin.

Although numerous experimental models of skin substitutes have been described, few have become part of the routine protocols for closure of skin wounds. To be adopted as a standard therapy, an experimental model must be compared to conventional therapies for determination of relative efficacy. Only if clinical advantages are realized does the experimental model become a new therapeutic standard. Determination of therapeutic advantages requires studies at both the preclinical and clinical levels. In this context, a collagen-glycosaminoglycan substrate populated with cultured fibroblasts and keratinocytes has demonstrated advantages over split-thickness skin grafts.

PRECLINICAL STUDIES OF GRAFT COMPOSITION AND WOUND CLOSURE

A diagram of an overall process for fabrication and application of skin substitutes is shown in Figure 4.1. Epidermal keratinocytes, melanocytes and dermal fibroblasts are isolated from a skin biopsy. For autologous skin substitutes, the biopsy may be a small (5-10 cm^2) sample of split-thickness skin as used for grafting of excised burns, or of full-thickness skin with primary closure of the site for treatment of chronic wounds. Conversely, a sample of allogeneic skin may be obtained from a cadaveric donor at an accredited tissue bank. Currently, preparation of allogeneic cell cultures from living donors (i.e., surgical discard) in the United States requires testing for pathogens (e.g., HIV, CMV, hepatitis) at least six months after donation before clearance for transplantation. Unless the living donor agrees to subsequent testing, compliance may not be expected. Therefore, the most dependable source of allogeneic skin is a cadaveric donor, and adaptation of safety standards used in tissue banking.[52]

After isolation, cells are propagated to very large populations in selective media.[53-55] Individual cell types may be cryopreserved by controlled rate freezing[56] in liquid nitrogen for virtually indefinite periods of time. As needed, keratinocytes and fibroblasts are recovered into culture at greater than 50% colony-forming efficiency.[57] After expansion into large populations, cells are inoculated at relatively high density (2.5-5.0 x 10^5 cells/cm^2) onto a biopolymer substrate. The collagen-GAG substrate described below has an asymmetric structure with one side porous, and one side micro-porous. Fibroblasts are inoculated first onto the porous side of the substrate, and incubated

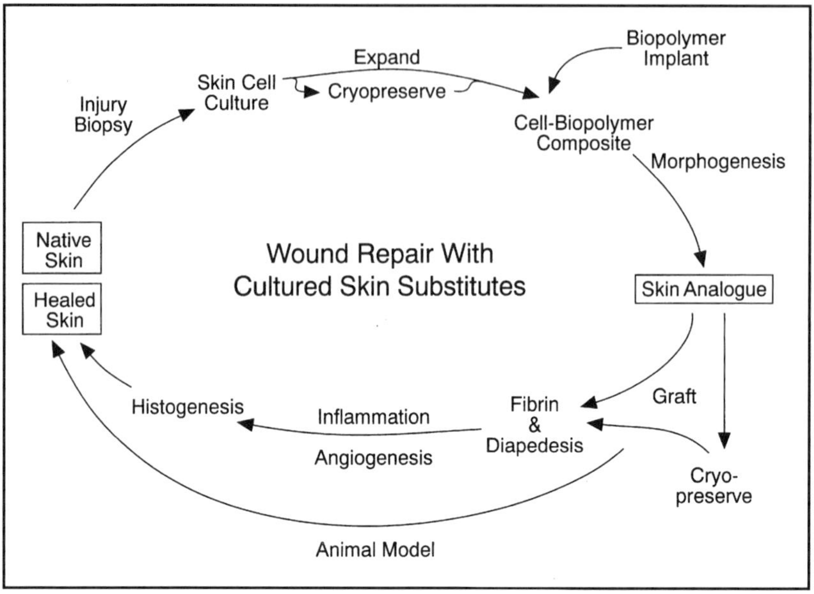

Fig. 4.1. Diagram of process for fabrication, testing and use of skin substitutes. Skin analogs result from "morphogenesis" of cultured skin cells in combination with biopolymer substrates. "Histogenesis" occurs only after grafting and healing of skin substitutes. "Organogenesis" does not occur.

overnight. The following day, the substrate is inverted and the keratinocytes are inoculated. On day 3 after inoculation of the keratinocytes, the cell-biopolymer material is air-exposed to promote epithelial stratification and barrier formation.[58-60] At this point, a skin analog is formed that may be compared anatomically and physiologically to native skin. At present, no cultured skin substitute has duplicated native skin, none is stable in culture for extended periods of time, and none survives cryopreservation without substantial loss of cellular viability.

After grafting to full-thickness wounds, a skin analog becomes revascularized, the biopolymer implant degrades, transplanted cells reform functional tissue during the inflammatory and proliferative phases of wound healing, and stability of tissue is restored during the resolution and remodeling phases of wound healing. At this point, skin substitutes regain homology with native skin with respect to barrier function, mechanical strength and perhaps pigmentation. A current objective is to obtain the degree of tissue integrity and stability in vitro that occurs after wound healing in vivo.

In Vitro Studies

Fabrication of cultured skin substitutes that are applied in a single procedure requires a delivery vehicle for skin cells that will ultimately reform skin tissue. Resolution of the healing process requires elimination of the polymer from the wound. To satisfy requirements of biocompatibility, implants must be degraded in the healing wound and replaced with native biopolymers synthesized by transplanted and resident cells. Therefore, the biopolymer implant must be compatible both to the transplanted cells and to the healing wound.

Figure 4.2 shows a diagram for general fabrication of collagen-carbohydrate substrates. These general procedures are similar to those described by Yannas[33] with certain modifications. Medical grade bovine type I/type III dermal or tendon collagen and GAG (i.e., chondroitin-sulfate, hyaluronic acid, dermatan sulfate) are solubilized in 0.5 M acetic acid. The two solutions are mixed in a refrigerated stirrer by slow addition of the GAG to the collagen to form a coprecipitate that is blended to homogeneity. The aqueous coprecipitate is cast and frozen rapidly to maximize the freezing rate. Because the ice crystal size and pore size in the lyophilized substrate are inversely proportional to rate of freezing, rapid freezing produces smaller pores.[61] After freezing is complete, the frozen liquid is lyophilized overnight, and then cross-linked by thermal dehydration in a vacuum oven at 105°C for 24 hours. The dry substrate has a sponge structure with pore diameters ranging from 50-150 μm, and an approximate dry weight of 5 mg/cm^2. A micro-porous film of the same collagen-GAG is added to one side of the sponge by addition of 3% v/v dimethyl sulfoxide to the coprecipitate. A liquid film is prepared and contacted to the dry sponge followed by rapid freezing, freeze-drying and thermal dehydration. The laminated sponge is packaged in a medical peel-pack, sterilized by gamma-irradiation, and stored until needed. At this point in the process, substrates can be stored for later use which is definitive for increased availability of skin substitutes. As needed, sterile substrates are rehydrated and inoculated with cells as described above. Rehydrated substrates may be stored before inoculation of cells for up to 30 days in 70% isopropanol with washing in culture medium before use. An important distinction of this process from Yannas is the elimination of glutaraldehyde cross-linking of the substrate that inhibits its degradation in the wound.[62]

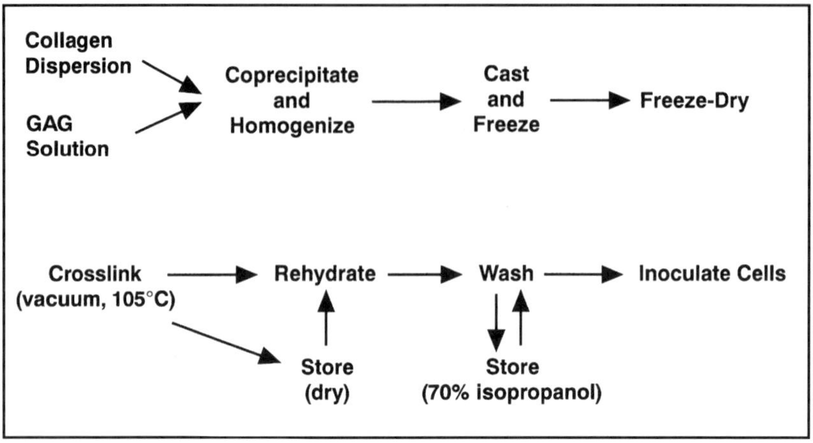

Fig. 4.2. Diagram of fabrication process for collagen-glycosaminoglycan (GAG) substrates.

Restoration of epidermal barrier in vitro from cultured keratinocytes requires regulation of proliferation and differentiation. Accomplishment of this objective depends on providing conditions in which the complex lipids that constitute epidermal barrier can be synthesized, released and organized in the intercorneocyte space. Figure 4.3 shows a diagram of relationships among pathways of carbohydrate-lipid metabolism in epidermis. Barrier is one of several endpoints for metabolism all of which begin with glucose uptake by keratinocytes. Besides energy production, other major endpoints include glycogen, cell membrane and storage lipid as triglyceride and sterol esters. Lipid analysis of epidermal analogs and native human skin after incubation in conventional culture media has demonstrated great excesses of storage lipid and great deficiencies of barrier lipids.[63,64] These imbalances can be attributed to deficiencies in culture conditions, and particularly in formulations of nutrient media. In part, these metabolic imbalances result from use of medium formulations that promote proliferation for the purpose of stimulating differentiated function. For example, it is known that epidermal growth factor (EGF) is a potent mitogen for proliferation of keratinocytes.[57] However, the fraction of S-phase keratinocytes in skin substitutes increases if EGF is removed from media.[65] This biphasic activity of EGF results from reduction in the mitotic rate of confluent keratinocytes by density inhibition.[66] Conversely, concentrations of autocrine growth factors (e.g., transforming growth factor-α; insulin-like growth factor-1) increases

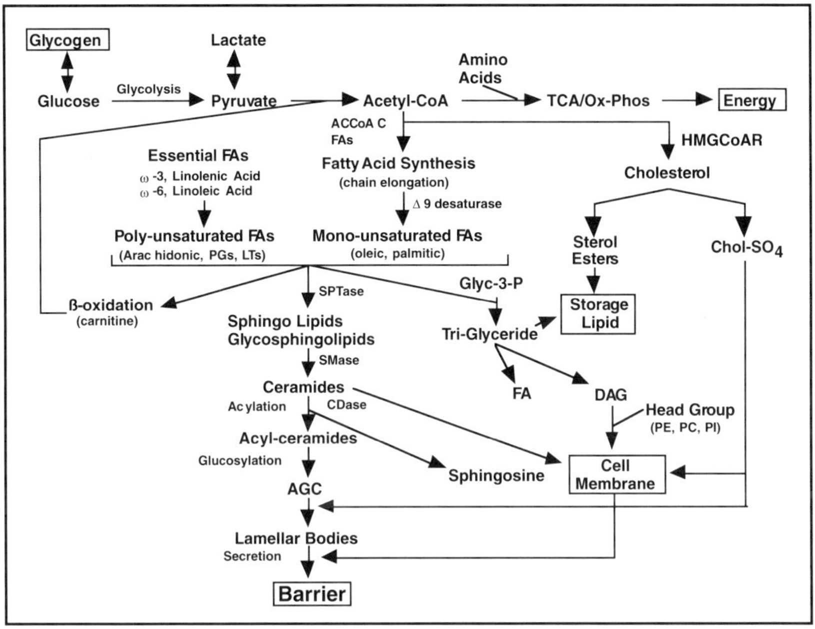

Fig. 4.3. Diagram of carbohydrate-lipid metabolism. Excessive accumulation of storage lipid, and deficiencies of barrier lipids suggest that epidermal cultures are hyperproliferative in most nutrient media.

proportionately to keratinocyte density to reduce requirements for exogenous addition of peptide growth factors. Therefore, it may be expected that improved barrier formation in keratinocyte cultures will result from reduction both of exogenous addition of mitogens, insulin, and energy substrates such as glucose and glutamine. Recent studies[67,68] in which these factors have been decreased or eliminated in maturing epidermal analogs have demonstrated more functional and stable barrier in skin substitutes.

Another historic principle of cell culture is the selection of a single cell type by the formulation of a culture medium. This principle has resulted in a wide variety of normal human cell types to be propagated in culture in low-serum, serum-free or biochemically defined media.[69] However, this basic principle does not always apply for cocultures of two or more cell types as in skin substitutes. To preserve the potential of skin substitutes to heal, priority must be placed on barrier formation by keratinocytes. But duplication of native physiology must also allow survival of fibroblasts, melanocytes, endothelial and immune cells, nerve and smooth muscle. Reestablishment of native physiology would permit expression of native phenotypes by all cells in skin substitutes. Accomplishment of this complex and dynamic equilibrium remains a continuing challenge in preparation of cultured skin substitutes. The histologic appearance of a cultured skin substitute with a stratified epidermal component, and a collagen-GAG-fibroblast dermal substitute is shown in Figure 4.4. Vascular plexus, nerves, glands, follicles, and immune effector cells are absent. Furthermore, the perishable nature of skin substitutes in vitro confirms the remaining deficiencies of current culture systems.

Animal Studies

Stable restoration of definitive functions of skin (barrier, pigment, mechanical strength) after transplantation of skin substitutes to wounds may be termed "histogenesis." Because preclinical studies of grafting involve human skin substitute, the athymic mouse has been used by several laboratories.[70-73] Figure 4.5 shows a photograph of an athymic mouse (Fig. 4.5A) with pigmented skin generated by healing of a full-thickness skin wound with a cultured skin substitute and a representative histology (Fig. 4.5B). Healed skin is smooth and without

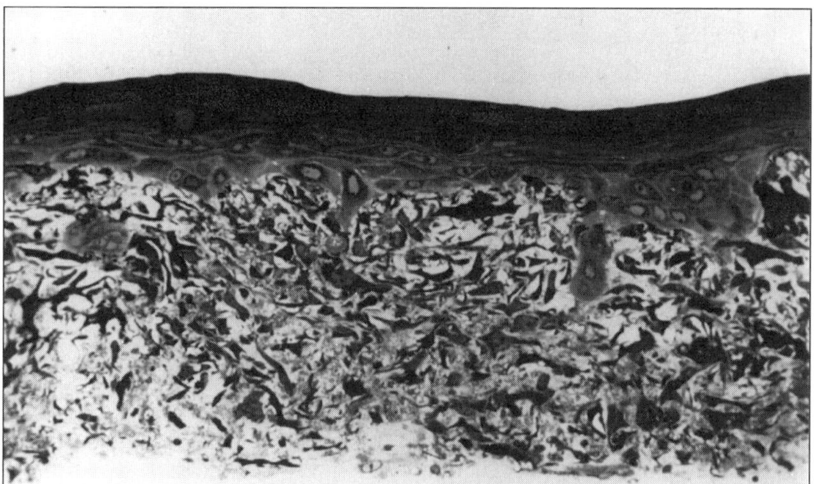

Fig. 4.4. Photomicrograph of a cultured skin substitute. Cultured human keratinocytes and fibroblasts attach to and proliferate on the collagen-GAG substrate. The epidermal component forms an analog of stratum corneum produced by dividing keratinocytes. Fibroblasts migrate and proliferate to fill the pores of the collagen-GAG. Scale bar = 0.1 mm.

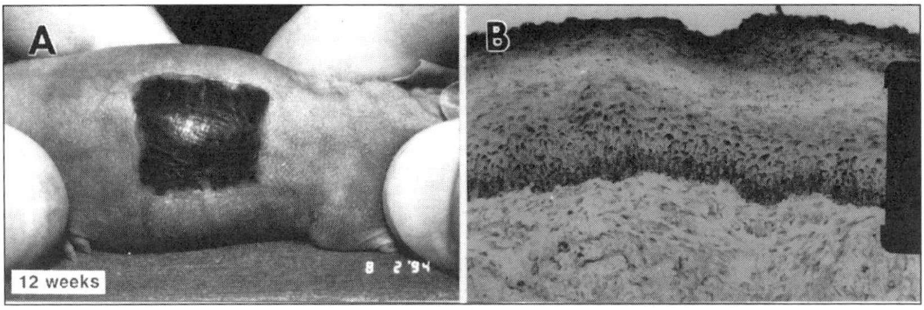

Fig. 4.5. Preclinical model of wound healing and pigmentation with cultured skin substitutes (CSS). (A) Athymic mouse at 12 weeks after grafting of CSS. Healed human skin is smooth, soft and uniformly pigmented. (B) Histology of pigmented skin after grafting of CSS containing melanocytes shows high levels of melanin produced by melanocytes that organize in the basal layer of keratinocytes. Pigment is transferred to keratinocytes and distributed throughout the epidermis. Scale bar = 0.25 mm.

glands or follicles. Pigment in skin substitutes results from "passenger melanocytes,"[71,74] or can be regulated by addition of cultured melanocytes to skin substitutes.[75] After healing, melanocytes organize in the basal layer of keratinocytes and transfer pigment to keratinocytes in all epidermal strata.

Other quantitative endpoints for these studies have included, but are not limited to: wound area,[75,76] ordinal scoring of human cell markers,[77] expression of specific proteins, mRNAs, and lipids.[34,78,79] Particular attention has been given to detection and quantification of factors for angiogenesis and proliferation (i.e., basic fibroblast growth factor, transforming growth factors-α & β, platelet-derived growth factor, insulin-like growth factors), inflammatory mediators (i.e., interleukins, arachidonic acid metabolites), and extracellular matrix components (i.e., collagens, polysaccharides, laminin, kalinin) by transplanted cells.[80-83] Together, the qualitative and quantitative proportions of cellular and extracellular factors define tissue integrity.

Assessments of efficacy of cultured skin substitutes in preclinical models have certain strengths and limitations. Rates of engraftment and wound area can be determined with reasonable validity. But because rodents do not scar extensively, the formation of keloid or hypertrophic scars cannot be evaluated well. However, recovery of skin function can be evaluated with biophysical instrumentation[84] as represented in Figure 4.6. Recovery of epidermal barrier can be assessed by surface electrical capacitance (SEC)[72] as shown in Figure 4.6A. During culture, skin substitutes develop partial barrier as represented

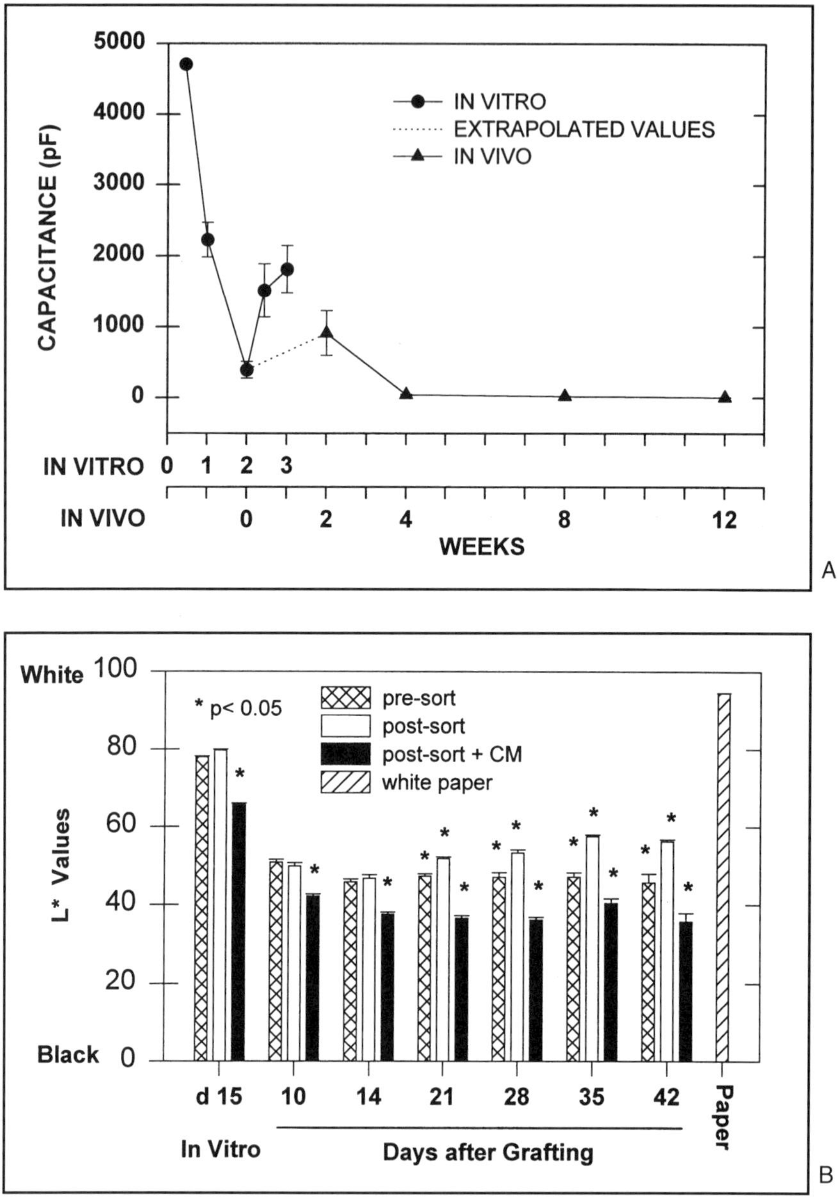

Fig. 4.6. Data plots of biophysical measurements. (A) Surface Electrical Capacitance is an accurate and convenient measurement of epidermal barrier. Complete and stable barrier forms only after engraftment onto wounds. Reprinted with permission from Boyce et al, J Invest Dermatol 1996; 107(1):82-87. (B) Skin pigmentation is measured with the Minolta Chromameter with L* (black-white) values as one axis of a three dimensional color space. Addition of melanocytes to CSS (post-sort + CM) results in healed skin that is darker statistically than control CSS (post-sort).

by decrease in SEC, but subsequent increase occurs after two weeks incubation. However, grafting to mice results in complete and stable formation in barrier as represented by very low SEC values. Similarly, redevelopment of pigment in grafted skin substitutes can be measured by the Minolta Chromameter as shown in Figure 4.6B. In this example, a quantitative and significant difference in pigmentation is identified by selective removal of melanocytes (post-sort) from keratinocyte cultures by flow cytometric sorting, and subsequent readdition of melanocytes (post-sort + CM) to generate black skin.[85] Hypothetically, a collection of instruments for measurement of biophysical properties of skin may allow objective and quantitative assessment of wound healing with skin substitutes in place of subjective clinical examinations.

CLINICAL STUDIES

Clinical Considerations

Anatomic and physiologic deficiencies in epidermal barrier and vascular supply confer on skin substitutes practical limitations that must be managed clinically to accomplish efficacy of wound closure. Most of these limitations are summarized in Table 4.2. Early reports with clinical application of keratinocyte sheets proved the life-saving potential of cultured skin substitutes.[86,87] These studies also identified multiple limitations to their routine use according to procedures for split-thickness skin. If grafted onto subcutaneous fat after excision of full-thickness burns, keratinocyte sheets attach poorly and are destroyed by mechanical shear and microbial contamination. To improve engraftment, proponents recommended during the 1980s excision to muscle fascia that resulted in removal of viable tissue and

Table 4.2. Clinical limitations and considerations for use of skin substitutes

Limitation	Consideration
Mechanical fragility	Special dressings and nursing care
Susceptibility to microbial contamination	Non-cytotoxic topical agents
Decreased rates of engraftment	Increased regrafting
Increased time to heal	Delay of recovery
Very high cost	Resource allocation

disfigurement. It was also noted that healed epithelium subsequently ulcerated and often required regrafting. These results increased rather than decreased time to completion of healing. In addition, the cost of preparation of keratinocyte sheets for large burns is extremely high. Later studies combined keratinocyte sheets with cadaveric allograft from which the allogeneic epidermis had been surgically removed.[29] This approach has gained more common use and is reported to have fewer complications.[30] However, it requires availability of viable cadaveric allograft skin.

Finally, the high cost of cultured skin substitutes remains a practical factor for clinical use. Estimates of cost of keratinocyte sheets range from $1,000-$13,000/% body surface area covered.[88,89] Those costs can approximately double if a dermal substitute is also used.[28,32] Therefore, costs can become limiting for treatment of large (60-90%) total body surface area burns with cultured skin substitutes. At present, no studies have convincingly demonstrated a savings of total hospitalization costs by use of cultured skin substitutes of any kind. For contemporary treatment of burns, cultured skin substitutes remain an important adjunct to conventional skin grafting[90] but are not a primary modality of wound closure except in extreme cases.

Qualitative Outcome

Combination of epidermal cells with dermal substitutes in vitro has been shown to virtually eliminate epidermal blistering after engraftment of transplanted keratinocytes.[22] More rapid formation of basement membrane and anchoring fibrils[91,92] in dermal-epidermal models may result from initiation of cell-substrate attachments in vitro and elimination of enzymatic release of cell sheets immediately before grafting. An important limitation of composite skin substitutes is lack of vascular supply in the dermal substitute. This limitation extends time for reperfusion, ischemia, and nutrient deprivation to transplanted cells that reduce rates of engraftment. Cell survival and engraftment may be improved significantly by irrigation of grafts with solutions of nutrients[93] and noncytotoxic antimicrobial agents.[94-96] This approach optimizes the wound environment to promote survival of transplanted cells and the management microbial contamination until vascularization is complete.

Collagen-GAG substrates populated with fibroblasts and keratinocytes have rates of clinical engraftment at two weeks after surgery that are approximately 60% compared with about 90% for skin autografts as shown in Figure 4.7. This amount of wound closure is meaningful in cases of large burns, and is likely to increase as composition of cultured skin substitutes continues to improve. Rates of epithelial healing with this model of skin substitute is no different than with meshed autograft as shown in Figure 4.8. Surface electrical capacitance demonstrates that the epithelial surface of newly epithelialized wounds is dry by 12-14 days after grafting and remains stable thereafter. This definitive property of wound closure is not statistically different between cultured skin substitutes and split-thickness

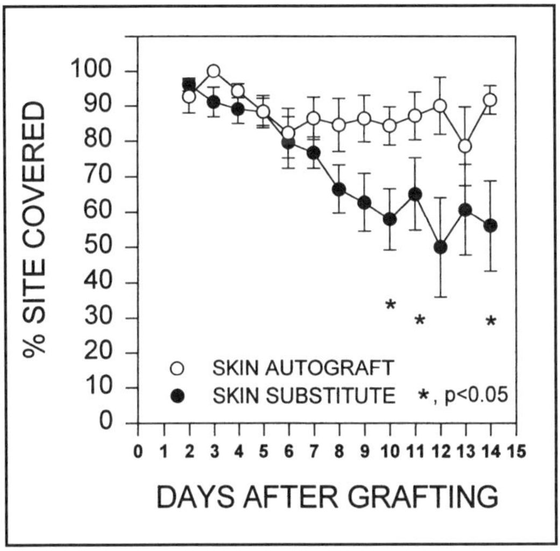

Fig. 4.7. Clinical wound coverage by cultured skin substitutes and skin autograft. Skin substitutes cover about 60% of wounds at two weeks after surgery, compared to about 90% for autograft. Reprinted with permission from Boyce et al, Ann Surg 1995; 222(6): 743-752.

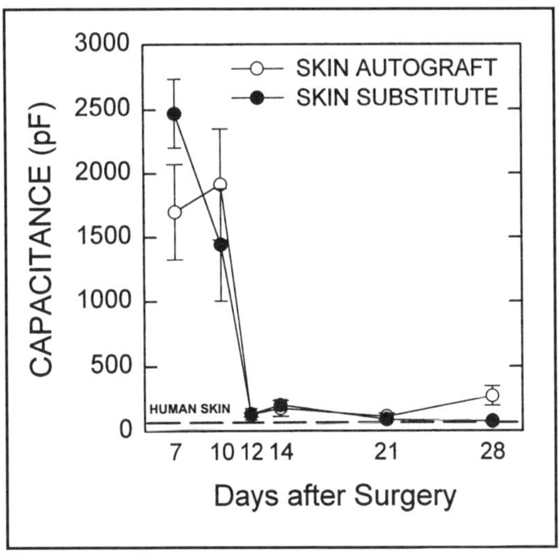

Fig. 4.8. Clinical formation of epidermal barrier by cultured skin and skin autograft. Wet dressing are discontinued at day 10 after surgery, and complete barrier forms within two days in both kinds of grafts. Reprinted with permission from Goretsky et al, Wound Repair Regen 1995; 3(4): 419-425.

skin autografts. Clinical repigmentation of skin substitutes has also been reported,[97] and is demonstrated qualitatively in Figure 4.9. Two outcomes with pigmentation of skin substitutes are consistent. First, pigmentation develops after epithelialization is complete. Skin barrier in composite grafts is stable by 2-3 weeks after surgery, but pigmentation begins to develop at 3-4 weeks and is usually not complete until 4-8 weeks. Second, skin substitutes are more consistently hyperpigmented than split-thickness skin autografts. This result is represented quantitatively in Figure 4.10. Pigmentation proceeds between 1-6 months as discreet foci of pigment within a hypopigmented epithelium (Fig. 4.10B). At present, pigment derives from "passenger" melanocytes that survive in cultures of keratinocytes from which skin substitutes are prepared. Therefore, pigmentation is most often distributed in an irregular pattern initially, but eventually becomes confluent. Hyperpigmentation in cultured skin grafts is observed clinically for the first 6 months after grafting and then gradually resolves to approximately the native color of the autologous donor.

Regulatory Considerations

With appropriate management of clinical limitations, wound closure with cultured skin substitutes has been demonstrated for acute and chronic wounds.[17,35,98,99] Cultured skin substitutes are part of a spectrum of materials extending from cadaveric skin allograft on one end, to tissue-derived materials (keratinocyte sheets, dermal matrix), to specific combinations of cells and purified polymers at the other end. Because some kinds of skin substitutes, including cultured keratinocyte sheets, were considered initially a form of tissue banking, requirements for demonstration of efficacy were virtually absent.[100] Conversely, cultured skin substitutes consisting of cultured cells attached to biodegradable polymers are currently regulated in the United States as Class III (significant risk) medical devices with rigorous requirements for demonstration of safety and efficacy, and compliance with Good Manufacturing Practices (GMPs).

Regulatory complexities among these materials have been recognized by the US Food and Drug Administration (FDA) and the European Committee for Standardization (CEN). FDA has responded by assembly of an Inter-Center Focus Group on Wound Healing to determine whether uniform standards for evaluation of safety and efficacy can be established for wound treatment materials.[101] This

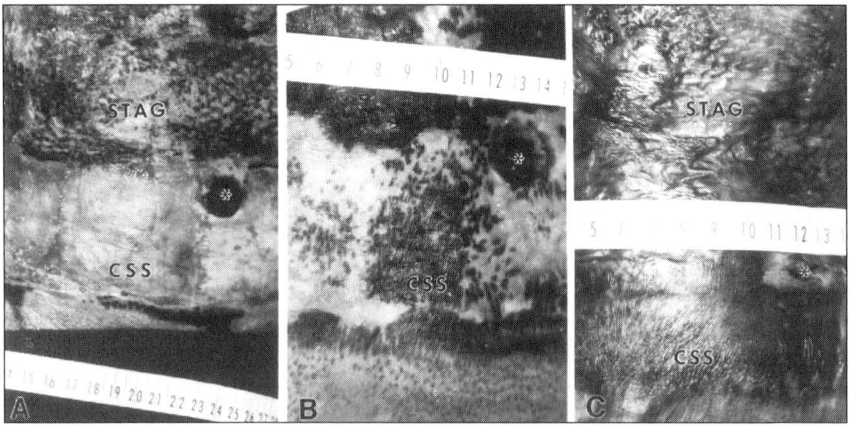

Fig. 4.9. Clinical repigmentation of cultured skin substitutes (CSS) and split-thickness autograft (STAG). The umbilicus of the patient is indicated (*) for orientation. (A) By day 10 after surgery, CSS is epithelialized and protects the wound, but has no pigment. (B) After 1 month, multiple foci of pigment develop within the CSS site. (C) By 4 months after grafting, the healed CSS is more uniformly pigmented and smoother than meshed, split-thickness autograft (STAG).

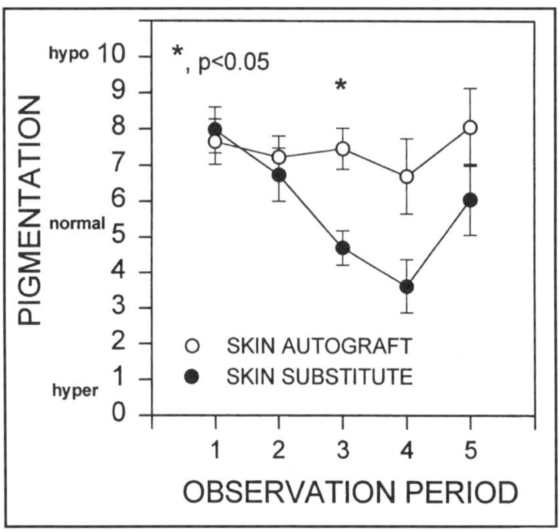

Fig. 4.10. Clinical hyperpigmentation of CSS after grafting is greater than skin autograft. During the first 6 months after grafting, CSS become progressively more pigmented, but hyperpigmentation fades slowly back to the native color of the patient's uninjured skin. Observation periods are: 1, 14-30 days; 2, 1-2 months; 3, 3-4 months; 4, 5-12 months; 5, 1 year and after. Reprinted with permission from Boyce et al, Ann Surg 1995; 222(6): 743-752.

ambitious goal may provide at the minimum a set of reference end points, and at the maximum an absolute set of parameters and methods for evaluation of experimental therapies including tissue-engineered skin. Similarly, CEN has established three directives pertaining to medical devices: the medical devices directive, the active implantable medical devices directive, and the in vitro diagnostic medical devices directive.[102] The CEN adopted its directives in this decade as part of the process of harmonization within the emerging European Community. Many parallels exist between the FDA and CEN standards, for example, in classification of medical devices. As economic globalization proceeds, resources will be conserved if regulatory standards can be harmonized concurrently. The International Standards Organization (ISO) has established regulations for control of sterility in product manufacture that overlap with standards for GMPs.[103,104] Because the underlying factors for medical safety and efficacy supersede national citizenship, progressive globalization of standards for medical devices that include cultured cells may prospectively be addressed by the World Health Organization or a similar affiliation of international experts. Current standards in the US recognize generally that wound closure (re-epithelialization) is an acceptable end point of wound healing with cultured skin substitutes. However, a distinction can readily be made between percent engraftment, and percent coverage of the body surface because in burns, percent engraftment of cultured skin is inversely related to magnitude of injury.[105] Standardization of these kinds of end points and distinctions in assessment will benefit all patients who are treated with cultured skin substitutes.

FUTURE DIRECTIONS

As anatomy and physiology of cultured skin substitutes improves, they will become more homologous to native skin graft. Improvement of skin substitutes will result from modifications of culture media, substrates and physical environment (i.e., humidity, mechanical tension, electrical properties) that have more fidelity to native skin. Better homology may be expected to reduce stringency for clinical use of cultured skin substitutes and accomplish the efficacy of skin autograft. After predictable efficacy is shown with autologous cells, successful models will become platforms for testing of chimeric grafts and genetically modified cells.[82,83,106,107] Gene therapy for treatment of local or systemic conditions is feasible with cultured skin substitutes.[82,106,108-111] For wound healing applications, transient expression of selected gene products may be best, whereas, constitutive gene expression may be required for systemic deficiencies. Delivery of gene products systemically probably will require physiologic regulation to be efficacious.[46]

If engineering of skin and other tissues is to become a true science, then members of its community must begin to think and act like engineers. Therefore, like other engineering disciplines, uniform standards for quantitative analysis of cultured skin substitutes must be established for evaluation of materials composition and performance. This goal will require interdisciplinary review and development of consensus. For this purpose, instrument-based, noninvasive assessment of outcome will be as important as materials analysis. Instrumental assessments may include measurement of epidermal barrier,[72,112] blood flow,[113] pigmentation and erythema,[114,115] visco-elastic properties,[116,117] and surface texture.[118] Establishment of these engineering standards will provide a platform from which validation and introduction of skin and other tissue substitutes will be expedited. Although the term "tissue engineering" describes a relatively new interdisciplinary field, it has its roots many years past in the fields of cell culture and polymer chemistry.[119-121] If past progress in engineering of cultured skin substitutes is an indication of progress, it is easy to predict the reduction of medical morbidity and mortality by use of cultured skin substitutes.

Acknowledgments

The author's studies are supported by National Institutes of Health grant GM 50509, Food and Drug Administration grant FD-R-672, and grants from the Shriners Hospitals for Children.

References

1. Burns, causes and costs: a Burn Foundation Report. Philadephia, PA: 1989.
2. Meehan M. Multisite pressure ulcer prevalence survey. Decubitus 1990; 3(4):14-17.
3. Miller H, Delozier J. Cost implications of the pressure ulcer treatment guideline. Columbia, MD: Center for Health Policy Studies; Contract 282-91-0070, 1994.
4. The skin. In: Guides to evaluation of permanent impairment. American Medical Association, 1993:277-289.
5. Tanner JC, Vandeput J, Olley JF. The mesh skin autograft. Plast Reconstr Surg 1964; 34:287-292.
6. Alexander JW, MacMillan BG, Law E, Kittur DS. Treatment of severe burns with widely meshed skin autograft, and meshed skin allograft overlay. J Trauma 1981; 21(6):433-438.

7. Housinger TA, Warden GD, Lang D, Visse M, Pohlman S. The use of sheet skin grafts in pediatric burn patients. Proc Amer Burn Assn 1991; 23:32. (Abstract).
8. Housinger TA, Hills J, Warden GD. Management of pediatric facial burns. J Burn Care Rehabil 1994; 15(5):408-411.
9. McCauley RL, Oliphant JR, Robson MC. Tissue expansion in the correction of burn alopecia: classification and methods of correction. Ann Plast Surg 1990; 25(2):103-115.
10. Neale HW, High RM, Billmire DA, Carey JP, Smith D, Warden GD. Complications of controlled tissue expansion in the pediatric burn patient. Plast Reconstr Surg 1988; 82(5):840-848.
11. Alghanem AA, McCauley RL, Robson MC, Rutan RL, Herndon DN. Managment of pediatric perineal and genital burns: twenty-year review. J Burn Care Rehabil 1990; 11(4):308-311.
12. Hendricks WM, Swallow RT. Management of stasis leg ulcers with Unna's boot versus elastic support stockings. J Am Acad Dermatol 1985; 12(1):90-98.
13. Bergstrom N, et.al. Ulcer care. In: Clinical Practice Guideline, Number 15: Treatment of Pressure Ulcers. Anonymous, ed. Rockville, MD: U.S. Department of Health and Human Services, 1994:45-57.
14. Burr WM, Elliott BM, Robison JG, Brothers TE. The etiology and treatment of venous stasis ulcers. J South Carolina Med Assoc 1993; 89(2):67-70.
15. Jamieson WG, DeRose G, Harris KA. Management of venous stasis ulcers: long-term follow-up. Canadian J Surg 1990; 33(3):222-223.
16. Phillips TJ, Bhawan J, Leigh IM, Green H, Gilchrest BA. Cultured epidermal allografts: clinical use, survival, and expression of maturation markers. Clin Research 1988; 36A:684
17. Leigh IM, Purkis PE, Navsaria HA, Phillips TJ. Treatment of chronic venous ulcers with sheets of cultured allogeneic keratinocytes. Br J Dermatol 1987; 117:591-597.
18. DeLuca M, Albanese E, Cancedda R, Viacava A, Faggioni A, Zambruno G, Gianetti A. Treatment of leg ulcers with cryopreserved allogeneic cultured epithelium: A multicenter study. Arch Dermatol 1992; 128:633-638.
19. Phillips TJ, Kehinde O, Green H, Gilchrest B. Treatment of skin ulcers with cultured epidermal allografts. J Am Acad Dermatol 1989; 21:191-199.
20. Aubock J, Irschick E, Ramani N et al. Rejection after slightly prolonged survival time of Langerhans call-free allogenic cultured epidermis used for wound coverage in humans. Transplantation 1988; 48:730-737.
21. Hansbrough JF. Current status of skin replacements for coverage of extensive burn wounds. J Trauma 1990; 30:S155-S160.
22. Hansbrough JF, Boyce ST, Cooper ML, Foreman TJ. Burn wound closure with cultured autologous keratinocytes and fibroblasts at-

tached to a collagen-glycosaminoglycan substrate. JAMA 1989; 262:2125-2130.
23. Rouabhia M, Germain L, Bergeron J, Auger FA. Successful transplantation of chimeric allogeneic-autologous cultured epithelium. Cell Transplant 1994; 3(3):224.
24. Rouabhia M, Germain L, Bergeron J, Auger FA. Allogeneic-syngeneic cultured epithelia: a successful therapeutic option for skin regeneration. Transplantation 1995; 59(9):1229-1235.
25. Heck EL, Bergstresser PR, Baxtor CR. Composite skin graft: Frozen dermal allografts support the engraftment and expansion of autologous epidermis. J Trauma 1985; 25(2):106-112.
26. Yannas IV, Burke JF, Orgill DP, Skrabut EM. Wound tissue can ustilize a polymeric template to synthesize a functional extension of skin. Science 1982; 215:174-176.
27. Burke JF, Yannas IV, Quinby WC, Bondoc CC, Jung WK. Successful use of a physiologically acceptable skin inthe treatment of extensive burn injury. Ann Surg 1981; 194:413-428.
28. Heimbach D, Luterman A, Burke JF, Cram A, Herndon D, Hunt J, Jordon M, McManus W, Solem L, Warden G, Zawacki B. Artifical dermis for major burns; a multi-center randomized clinical trial. Ann Surg 1988; 208:313-320.
29. Cuono C, Langdon R, Birchall N, Barttelbort S, McGuire J. Composite autologous-allogeneic skin replacement: development and clinical application. Plast Reconstr Surg 1987; 80:626-635.
30. Compton CC, Hickerson W, Nadire K, Press W. Acceleration of skin regeneration from cultured epithelial autografts by transplantation to homograft dermis. J Burn Care Rehabil 1993; 14(6):653-662.
31. Livesey SA, Herndon DN, Hollyoak MA, Atkinson YH, Nag A. Transplanted acellular allograft dermal matrix. Potential as a template for the reconstruction of viable dermis. Transplantation 1995; 60(1):1-9.
32. Wainwright D, Madden M, Luterman A, Hunt J, Monafo W, Heimbach D, Kagan R, Sittig K, Dimick A, Herndon D. Clinical evaluation of an acellular allograft dermal matrix in full-thickness burns. J Burn Care Rehabil 1996; 17:124-136.
33. Yannas IV, Burke JF. Design of an artifical skin. I Basic design principles. J Biomed Mater Res 1980; 14:65-81.
34. Boyce ST, Greenhalgh DG, Housinger TA, Kagan RJ, Rieman M, Childress CP, Warden GD. Skin anatomy and antigen expression after burn wound closure with composite grafts of cultured skin cells and biopolymers. Plast Reconstr Surg 1993; 91:632-641.
35. Boyce ST, Goretsky MJ, Greenhalgh DG, Kagan RJ, Rieman MT, Warden GD. Comparative assessment of cultured skin substitutes and native skin autograft for treatment of full-thickness burns. Ann Surg 1995; 222(6):743-752.

36. Hansbrough JF, Dore C, Hansbrough WB. Clinical trials of a living dermal tissue replacement placed beneath meshed, split-thickness skin grafts on excised wounds. J Burn Care Rehabil 1992; 13:519-529.
37. Hansbrough JF, Cooper ML, Cohen R, Spielvogel RL, Greenleaf G, Bartel RL, Naughton G. Evaluation of a biodegradable matrix containing cultured human fibroblasts as a dermal replacement beneath meshed skin grafts on athymic mice. Surgery 1992; 111(4):438-446.
38. Beumer GJ, van Blitterswijk CA, Ponec M. Biocompatibility of a biodegradable matrix used as a skin substitute: an in vivo evaluation. J Biomed Mater Res 1994; 28(5):545-552.
39. Beumer GJ, van Blitterswijk CA, Bakker D, Ponec M. Cell-seeding and vitro biocompatibility evaluation of polymeric matrices of PEO/PBT copolymers and PLLA. Biomaterials 1993; 14(8):598-604.
40. Madden MR, Finkelstein JL, Staiano-Coico L, Shires GT. Grafting of cultured allogenic epidermis on second-and third-degree burn wounds on 26 patients. J Trauma 1986; 26:955-962.
41. Madden MR, Finkelstein JL, Halbert C, Staiano-Coico L, Goodwin CW. Skin Substitutes: Current Clinical Status. Contemp Surg 1994; 45(1):23-26.
42. Roseeuw DI, de Coninck A, Lissens W, Kets E, Liebaers I, Vercruysse A, Vandenberge Y. Allogeneic cultured epidermal grafts heal chronic ulcers although they do not remain as proved by DNA analysis. J Dermatol Sci 1990; 1:245-252.
43. Phillips T, Provan A, Colbert D, Easley K. A randomized single-blind controlled study of cultured epidermal allografts in the treatment of split-thickness skin graft donor sites. Arch Dermatol 1993; 129: 879-882.
44. Sabolinski M, Rovee D, Parenteau N, Mulder G, Jensen J. The efficacy and safety of GraftSkin for the treatment of chronic venous ulcers. Wound Rep Reg 1995; 3(1):78 (Abstract).
45. Boyce ST. Skin repair with the cultured cells and biopolymers. In: Wise D, ed. Biomedical Applications. Totowa, New Jersey: Humana Press, Inc., 1996: 347-377.
46. Boyce ST. Epidermis as a secretory tissue. J Invest Dermatol 1994; (editorial)102:8-10.
47. Lafferty KJ, Bootes A, Dart G, Talmage DW. Effect of organ culture on the survival of thyroid allografts in mice. Transplantation 1976; 22:138-149.
48. Lafferty KJ, Prowse SJ, Simeonovic CJ. Immunobiology of tissue transplantation: A return to the passenger leukocyte concept. Annu Rev Immunol 1983; 1:143-173.
49. Hefton JM, Amberson JB, Biozes DG, Weksler ME. Loss of HLA-DR expression by human epidermal cells after growth in culture. J Invest Dermatol 1984; 83:48-50.
50. Morhenn VB, Benike CJ, Cox AJ, Charron DJ, Engleman EG. Cultured human epidermal cells do not synthesize HLA-DR. J Invest Dermatol 1982; 78:32-37.

51. Wickner NE, Huff JC, Norris DA, Boyce ST, Cary M, Kissinger M, Weston WL. Study of HLA-DR synthesis in cultured human kerantinocytes. J Invest Dermatol 1986; 87:559-564.
52. Linden JV, Favreau TJ. Professional standards in cell and tissue processing. Cell Transplant 1995; 4(5):441-446.
53. Boyce ST, Ham RG. Cultivation, frozen storage, and clonal growth of normal human epidermal keratinocytes in serum-free media. J Tiss Cult Meth 1985; 9:83-93.
54. Bettger WJ, Boyce ST, Walthall BJ, Ham RG. Rapid clonal growth and serial passage of human diploid fibroblasts in a lipid-enriched synthetic medium supplemented with epidermal growth factor, insulin, and dexamethasone. Proc Natl Acad Sci USA 1981; 78:5588-5592.
55. Medrano EE, Nordlund JJ. Successful culture of adult human melanocytes obtained from normal and vitiligo donors. J Invest Dermatol 1990; 95(4):441-445.
56. Coriell LL. Preservation, storage and shipment. In Jakoby WB, Pastan IR, eds. Methods in Enzymology: Volume LVIII, Cell Culture. New York, NY: Academic Press, 1979:29-35.
57. Boyce ST, Ham RG. Calcium-regulated differentiation of normal human epidermal keratinocytes in chemically defined clonal culture and serum-free serial culture. J Invest Dermatol 1983; 81 suppl 1:33s-40s.
58. Prunieras M, Regnier M, Woodley DT. Methods for cultivation of keratinocytes at the air-liquid interface.. J Invest Dermatol 1983; 81(1):28s-33s.
59. Regnier M, Caron D, Reichert U, Schaefer H. Barrier function of human skin and human reconstructed epidermis. J Pharmaceutical Sci 1993; 82(4):404-407.
60. Mak VHW, Cumpstone MB, Kennedy AH, Harmon CS, Guy RH, Potts RO. Barrier function of human keratinocyte cultures at the airliquid interface. J Invest Dermatol 1991; 96:323-327.
61. Boyce ST, Christianson DJ, Hansbrough JF. Structure of a collagen-GAG dermal skin substitute optimized for cultured human epidermal keratinocytes. J Biomed Mater Res 1988; 22:939-957.
62. Harriger MD, Supp AP, Warden GD, Boyce ST. Glutaraldehyde cross linking of collagen implants in vitro inhibits degradation in cultured skin substitutes grafted to athymic mice. J Biomed Mater Res 1996; in press
63. Ponec M. Lipid metabolism in cultured keratinocytes. Adv Lipid Res 1991; 24:83-118.
64. Boyce ST, Williams ML. Lipid supplemented medium induces lamellar bodies and precursors of barrier lipids in cultured analogues of human skin. J Invest Dermatol 1993; 101:180-184.
65. Boyce ST, Hoath SB, Wickett R, Williams ML. Loss of requirement for exogenous epidermal growth factor by cultured analogous of human skin. J Invest Dermatol 1993; 100(4):579 (Abstract).

66. Holley RW, Armour R, Baldwin JH. Density-dependent regulationof growth of BSC-1 cell in cell culture: control of growth by low molecular weight nutrients. Proc Natl Acad Sci U S A 1978; 75(1):339-341.
67. Boyce ST, Vicanova J, Bouwstra J et al. Biochemical and biophysical analyses of epidermal barrier formation in cultured skin substitutes grafted to athymic mice. J Invest Dermatol 1996; 106(4):916 (Abstract).
68. Boyce ST, Swope VB, Warden GD. Responses of keratinocyte proliferation to insulin-like growth factor-1 (IGF-1) and human growth hormone in serum-free medium. J Invest Dermatol 1995; 104(4):641 (Abstract).
69. Ham RG, McKeehan WL. Media and growth requirements. Methods Enzymol 1979; 58:44-93.
70. Banks-Schlagel S, Green H. Formation of epidermis by serially cultivated human epidermal cells transplanted as an epithelium to athymic mice. Transplantation 1980; 29:308-313.
71. Boyce ST, Foreman TJ, English KB, Stayner N, Cooper ML, Sakabu S, Hansbrough JF. Skin wound closure in athymic mice with cultured human cells, biopolymers, and growth factors. Surgery 1991; 110:866-876.
72. Boyce ST, Supp AP, Harriger MD, Pickens WL, Wickett RR, Hoath SB. Surface electrical capacitance as a non-invasive index of epidermal barrier in cultured skin substitutes in athymic mice. J Invest Dermatol 1996; 107(1):82-87.
73. Medalie DA, Eming SA, Tompkins RG, Yarmush ML, Krueger GG. Evaluation of human skin reconstituted from composite grafts of cultured keratinocytes and human acellular dermis transplanted to athymic mice. J Invest Dermatol 1996; 107(1):121-127.
74. DeLuca M, Franzi A, D'Anna F, Zicca A, Albanese E, Bondanza S, Cancedda R. Co-culture of human keratinocytes and melanocytes: differentiated melanocytes are physiologically organized in the basal layer of the cultured epithelium. Eur J Cell Biol 1988; 46:176-180.
75. Boyce ST, Medrano EE, Abdel-Malek ZA, Supp AP, Dodick JM, Nordlund JJ, Warden GD. Pigmentation and inhibition of wound contraction by cultured skin substitutes with adult melanocytes after transplantation to athymic mice. J Invest Dermatol 1993; 100:360-365.
76. Boyce ST, Glafkides MC, Foreman TJ, Hansbrough JF. Reduced wound contraction after grafting of full-thickness wounds with a collagen and chondroitin-6-sulfate (GAG) dermal skin substitute and coverage with Biobrane. J Burn Care Rehabil 1988; 94:364-370.
77. Griffey ES, Hueneke M, Sukkar S, Wainwright D, Livesey SA. Production of a human in vitro reconstituted skin and grafting to a nude mouse model. Wound Rep Reg 1995; 3(1):92 (Abstract).
78. Compton CC, Gill JM, Bradford DA, Regaur S, Gallico GG III, O'Connor NE. Skin regenerated from cultured epithelial autografts

on full-thickness burn wounds from 6 days to 5 years after grafting. Lab Invest 1989; 60:600-612.
79. Compton CC, Tong Y, Trookman N, Zhao H, Roy D, Press W. Transforming growth factor alpha gene expression in culture human keratinocytes is unaffected by cellular aging. Arch Dermatol 1995; 131(6):683-690.
80. Goretsky MJ, Harriger MD, Supp AP, Greenhalgh DG, Boyce ST. Expression of interleukin 1α, interleukin 6, and basic fibroblast growth factor by cultured skin substitutes before and after grafting to full-thickness wounds in athymic mice. J Trauma 1996; 40:894-900.
81. Sahuc F, Nakazawa K, Berthod F, Collombel C, Damour O. Mesenchymal-epithelial interactions regulate gene expression of type VII collagen and kalinin in keratinocytes and dermal-epidermal junction formation in a skin equivalent model. Wound Rep Reg 1996; 4(1):93-102.
82. Eming SA, Lee J, Snow RG, Tompkins RG, Yarmush ML, Morgan JR. Genetically modified himan epidermis overexpressing *PDGF-A* directs the development of a cellular and vascular connective tissue stroma when transplanted to athymic mice-Implications for the use of genetically modified keratinocytes to modulate dermal degeneration. J Invest Dermatol 1995; 105:756-763.
83. Eming SA, Snow RG, Yarmush ML, Morgan JR. Targeted expression of insulin-like growth factor to human keratinocytes: modification of the autocrine control of keratinocyte proliferation. J Invest Dermatol 1996; 107(1):113-120.
84. Serup J, Jemec GBE. Non-Invasive Methods and the Skin. CRC Press, Inc.: Boca Raton, FL, 1995.
85. Swope VB, Cornelius JR, Babcock GF, Boyce ST. Flow cytometric separation of epidermal keratinocytes and melanocytes. J Invest Dermatol 1996; 106(4):882 (Abstract).
86. O'Connor NE, Mulliken JB, Banks-Schlagel S, Kehinde O, Green H. Grafting of burns with cultured epithelium from autologous epidermal cells. Lancet 1981; 1(8211):75-81.
87. Gallico III GG, O'Connor NE, Compton CC, Kehinde O, Green H. Permanent coverage of large burn wounds with autologous cultured human epithelium. N Engl J Med 1984; 311:448-451.
88. Munster AM, Weiner SH, Spence RJ. Cultured epidermis for coverage of burn wounds: a single center experience. Ann Surg 1990; 211:676-680.
89. Rue LW, Cioffi WG, McManus WF, Pruitt BA. Wound closure and outcome in extensively burned patients treated with cultured autologous keratinocytes. J Trauma 1993; 34(5):662-667.
90. Desai MH, Mlakar JM, McCauley RL, Abdullah KM, Rutan RL, Waymack JP, Robson MC, Herndon DN. Lack of long term durability of cultured keratinocyte burn wound coverage: a case report. J Burn Care Rehabil 1991; 12:540-545.

91. Hull BE, Finley RK, Miller SF. Coverage of burns with bilayered skin equivalents: a preliminary clinical trial. Surgery 1990; 107:496-502.
92. Cooper ML, Andree C, Hansbrough JF, Zapata-Sirvent RL, Speilvogel RL. Direct comparison of a cultured composite skin substitute containing human keratinocytes and fibroblasts to an epidermal sheet graft containing human keratinocytes on athymic mice. J Invest Dermatol 1993; 101(6):811-819.
93. Boyce ST, Supp AP, Harriger MD, Greenhalgh DG, Warden GD. Topical nutrients promote engraftment and inhibit wound contraction of cultured skin substitutes in athymic mice. J Invest Dermatol 1995; 104(3):345-349.
94. Teepe RGC, Kreis RW, Koebrugge EJ, Kempenaar JA, Vloemans AF, Hermans RP, Boxa H, Dokter J, Hermans J, Ponec M, Vermeer BJ. The use of cultured autologous epidermis in the treatment of extensive burn wounds. J Trauma 1990; 30:269-275.
95. Boyce ST, Holder IA. Selection of topical antimicrobial agents for cultured skin for burns by combined assessment of cellular cytotoxicity and antimicrobial activity. Plast Reconstr Surg 1993; 92(4):493-500.
96. Boyce ST, Warden GD, Holder IA. Non-cytotoxic combinations of topical antimicrobial agents for use with cultured skin. Antimicrob Agents Chemother 1995; 39(6):1324-1328.
97. Harriger MD, Warden GD, Greenhalgh DG, Kagan RJ, Boyce ST. Pigmentation and microanatomy of skin regenerated from composite grafts of cultured cells and biopolymers applied to full-thickness burn wounds. Transplantation 1995; 59:702-707.
98. Boyce ST, Glatter R, Kitzmiller WJ. Treatment of chronic wounds with cultured cells and biopolymers. Wounds 1995; 7(1):24-29.
99. Phillips TJ, Bhawan J, Leigh IM, Baum HJ, Gilchrest BA. Cultured epidermal autografts and allografts: a study of differentiation and allograft survival. J Am Acad Dermatol 1990; 23:189-195.
100. Kessler DA, Siegel JP, Noguchi PD, Zoon KC, Feiden KL, Woodcock J. Regulation of somatic-cell therapy and gene therapy by the Food and Drug Administration. N Engl J Med 1993; 329(16):1169-1173.
101. US Food and Drug Administration. Responses from the Wound Healing Clinical Focus Group at the FDA to the Government Relations Committee of the Wound Healing Society. Scars and Stripes 1994; 4:5-12.
102. Fries RC, Graber MD. Designing medical devices for conformance with harmonized standards. Biomed Instr Technol 1995; 29:284-292.
103. Kauffman JM, Weiler ED. Application of ISO 9002 and FDA's good manufacturing practices to general chemical manufacturing. Quality Assurance 1992; 1(3):213-216.
104. Royal PD. Harmonization of good laboratory practice requirements and laboratory accreditation programs. Quality Assurance 1994; 3(3):312-315.

105. Williamson J, Snelling C, Clugston P, Mac Donald I, Germann E. Cultured epithelial autograft: Five years of clinical experience with twenty-eight patients. J Trauma 1995; 39:309-319.
106. Krueger GG, Morgan JR, Jorgensen CM, Schmidt L, Li HL, Li LT, Boyce ST, Wiley HS, Kaplan J, Peterson MJ. Genetically modified skin to treat disease: potentials and limitations. J Invest Dermatol 1994; 103(5):76s-84s.
107. Morgan JR, Barrandon Y, Green H, Mulligan RC. Expression of an exogenous growth hormone gene in transplantable human epidermal cells. Science 1987; 237:1476-1479.
108. Fenjves ES, Gordon DA, Pershing LK, Williams DL, Taichman LB. Systematic distribution of apolipoprotein E secreted by grafts of epidermal keratinocytes: implications for epidermal function and gene therapy. Proc Natl Acad Sci USA 1989; 86:8803-8807.
109. Vogt PM, Thompson S, Andree C, Liu P, Breuing K, Hatzis D, Brown H, Mulligan RC, Ericksson E. Genetically modified keratinocytes transplanted to wounds reconstitute the epidermis. Proc Natl Acad Sci U S A 1994; 91(20):9307-9311.
110. Fenjves ES, Smith J, Zaradic S, Taichman LB. Systemic delivery of secreted protein by grafts of epidermal keratinocytes: prospects for keratinocyte gene therapy. Hum Gene Ther 1994; 5(10):1241-1248.
111. Lu B, Scott G, Goldsmith LA. A model for keratinocyte gene therapy: preclinical and therapeutic considerations. Proc Assoc Amer Physicians 1996; 108(2):165-172.
112. Goretsky MJ, Supp AP, Greenhalgh DG, Warden GD, Boyce ST. Surface electrical capacitance as an index of epidermal barrier properties of composite skin substitutes and skin autografts. Wound Rep Reg 1995; 3(4):419-425.
113. Atiles L, Mileski W, Spann K, Purdue G, Hunt J, Baxter CR. Early assessment of pediatric burns by laser doppler flowmetry. J Burn Care Rehabil 1995; 16(6):596-601 (Abstract).
114. Feather JW, Hajizadeh-Saffiar M, Leslie G, Dawson JB. A portable scanning spectrophotometer using wavelenghts for the rapid measurements of skin pigments. Phys Med Biol 1989; 34:807-820.
115. Kollias N, Bager A. Quantitative assessment of UV-induced pigmentation and erythema. Photodermatol 1988; 5:53-60.
116. Agache P, Monneur C, Leveque J, DeRigal L. Mechanical properties and Youngs modulus of human skin in vivo. Arch Dermatol Res 1980; 269:221-232.
117. Matsuzaki K, Kumagai N, Fukushi S, Ohshima H, Tanabe M, Ishida H. Cultured epithelial autografting on meshed skin graft scars: evaluation of skin elasticity. J Burn Care Rehabil 1995; 16:496-502.
118. McQuiston B, Whitestone J. The application of laser surface scanning for quantifying human wound progression. Wound Rep Reg 1995; 3(1):78 (Abstract).

119. Hilleman MR. History, precedent, and progress in the development of mammalian cell culture systems for preparing vaccines: safety considerations revisited. J Med Virol 1990; 31(1):5-12.
120. Anderson JM, Gibbons DF. The new generation of bichemical polymers. Biomater Med Devices Artif Organs 1974; 2(3):235-248.
121. Lyman DJ. Biomedical polymers [Review]. Ann NY Acad Sci 1968; 146(1):30-48.

SECTION II

Experimental Use of Engineered Skin Substitutes

CHAPTER 5

Damaged Skin Replacement Using Tissue Engineering: Chimeric (Syngeneic-Allogeneic) Epidermal Substitute Production and Grafting

Mahmoud Rouabhia

INTRODUCTION

Skin is particularly vulnerable to environmental alterations. Although thermal burns are considered as the paradigm of severe skin injury, other conditions such as large congenital nevi, skin ulcers, nonhealing wounds and other types of traumatic cutaneous losses require therapeutic skin replacement.[1-3] Skin trauma and wounds occur at various degrees with distinct characteristics and intrinsic complications.[4-6] Replacing the damaged cutaneous tissue is the major challenge for the medical community. In such situations, the major objective is to replace defective skin as quickly as possible with high quality materials that contribute to cicatrization and aesthetic tissue aspect. The best replacement is without any doubt the patient's own skin (autografting).[7,8] However, in large and deeply burned patients, harvesting what is left as autologous healthy skin is dangerous and involves an additional trauma for each patient. Various bioengineered skin substitutes have been developed and used as permanent skin replacement or as temporary coverage until the occurrence of spontaneous or surgical cicatrization.

LARGE AND DEEP BURN PATIENT TREATMENT

Autologous Versus Allogeneic Epidermal Substitute Grafting

Burn injuries are examples of trauma that induce several defects which severely compromise vital defense mechanisms of patients.[9] The difficulty in the treatment of extensively and deeply (third and deep second degrees) burned patients is always the limited amount of donor sites available for autografting.

Engineered epidermal substitutes using the patient's keratinocytes have been developed and used as an adjunctive therapy for permanent skin replacement.[10,11] Considered as the only life-saving therapy for these patients, this autologous culture method needs several weeks for cell growth and graftable epidermal substitute production. This delay depends on the patient's age, the degree of thermal stress, the physiological suppressive factors released during the first hours of burn (major stress situation), etc.[12,13] Delay of several weeks is one of the major disadvantages for this therapeutic procedure. To overcome this handicap, engineered allogeneic epidermal substitutes have been suggested as functional damaged skin replacement for excessively burned patients. Indeed, the production and grafting of allogeneic epidermal substitutes allow permanent skin replacement.[14-16] The nonrejection of allogeneic cultured substitutes in burn patients, and extended survival of these grafts in several animal models, is largely due to the limited antigenicity of these grafts. The decreased immunogenicity of allogeneic keratinocytes is based upon the observations that:

1. keratinocytes do not express major histocompatibility complex (MHC) class II antigens
2. keratinocyte cultures are depleted of Langerhans cells
3. keratinocyte cultures are depleted of passenger leukocytes
4. keratinocyte cultures do not elicit a humoral response

This exciting approach promised complete treatment of extensively burned patients. Further experimental and clinical results obtained after cultured epidermal allograft, however, strongly suggest that allogeneic keratinocytes are rejected, and these epidermal substitute allografts are more immunogeneic than previously believed.[17-19] In unburned animal hosts, cultured epidermal allografts sensitize for accelerated second-set rejection as effectively as full-thickness allografts. This sensitization is associated with an enhanced cytotoxic lymphocyte response.[20] It was shown that IFN-γ stimulated keratinocytes express MHC class II antigens that can maintain a cytotoxic T cell clone involved in the allograft rejection.[21] Also, keratinocytes are potent producers of active cytokines.[22,23] These experimental and clinical data show that:

1. skin substitute autografting takes permanently
2. foreign skin substitute grafts are completely rejected
3. burn patients have to be covered quickly and permanently to keep them alive.

The question was if we can successfully and permanently replace damaged skin by mixing autologous (1) and foreign (2) keratinocytes (Fig. 5.1)? To answer this question we devised in an experimental model a new culture technology and produced chimeric graftable epidermal substitutes.

CHIMERIC CULTURE METHOD

Epidermal Cell Isolation and Culture

Donor cells are Balb/c and C3H/HeN mice. Each of these strains possesses completely different MHC. Indeed, for Balb/c the MHC

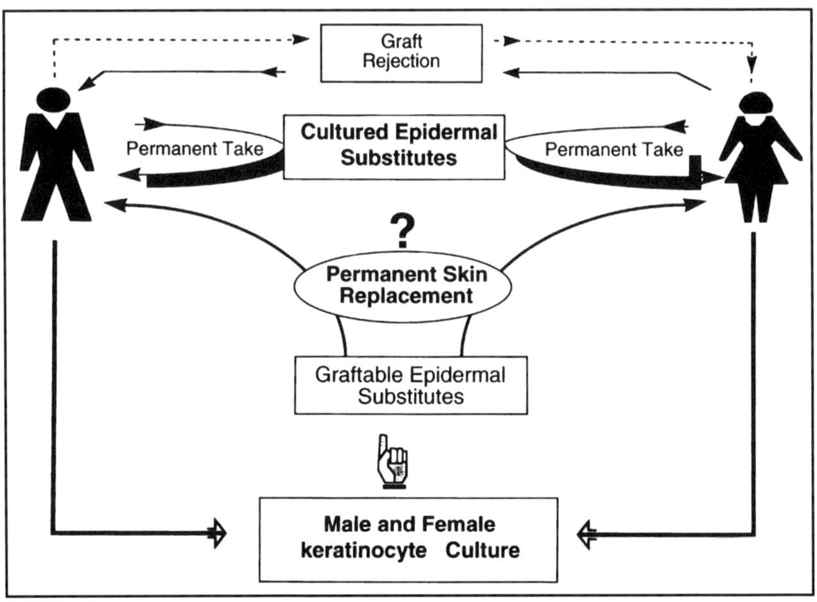

Fig. 5.1. Diagram on the well known state of skin and cultured epidermal substitute grafting.

class I is $H2^d$ compared to $H2^k$ for C3H/HeN, and their MHC class II is IA^d and IA^k, respectively. Epidermal cell suspensions were prepared from the whole skin of the two strains of newborn mice (1 to 3 days old); following the previously described method.[17,24] Both types of keratinocyte populations were plated separately to obtain homogeneous (100%) Balb/c and C3H/HeN keratinocyte seeding or cocultured as showed in the Table 5.1.

Cells were maintained in culture as previously described.[17,24] They were observed daily using an inverted microscope and photographed. These observations showed that keratinocytes in all chimeric cultures grow as well as homogeneous keratinocyte cultures (Fig. 5.2). Chimeric and homogeneous cultures reached confluence at the same time. Are both keratinocyte types still present in the confluent chimeric cultures? Is there an overgrowth of one cell type on the other, in the chimeric culture, or not?

Keratinocyte Phenotyping Prior to Chimeric Epidermal Substitute Grafting

The overgrowth of a particular cell type on the other during long-term culture has been reported, particularly with fibroblasts and keratinocytes. This phenomenon is linked to the intrinsic capacity of the cells to proliferate under culture conditions defined by the types of culture substratum, nutrients and growth factors.[25-27] This overgrowth phenomenon may occur in chimeric cell cultures. To evaluate this situation and answer the previously asked questions, confluent chimeric cultures were dissociated and cell suspensions were stained

Table 5.1. In vitro production of homogeneous and chimeric epidermal cultures for in vitro analyses

Cultures	Keratinocyte Donors	Final Percentage
Homogeneous		
1	Only Balb/c	100%B
2	Only C3H/HeN	100%C
Chimeric		
1	1/2 Balb/c + 1/2 C3H/HeN	50%B-50%C
2	1/4 Balb/c + 3/4 C3H/HeN	25%B-75%C
3	3/4 Balb/c + 1/4 C3H/HeN	75%B-25%C

C: C3H/HeN; B: Balb/c

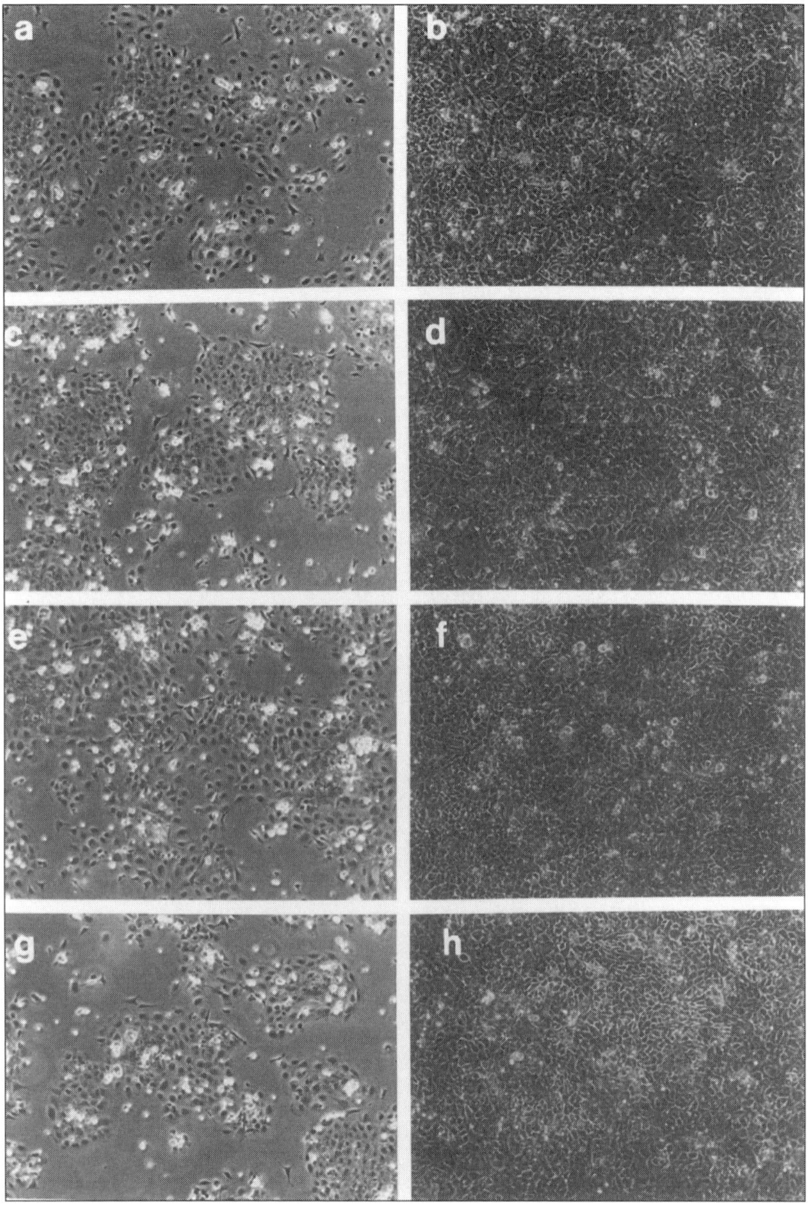

Fig. 5.2. Keratinocyte morphology. Epidermal cells were isolated from two different strains (Balb/c and C3H/HeN) and seeded separately or cocultured at different ratios: 50%-50% and 25%-75%. (a,c,e,g): 24 hours after seeding; (b,d,f,h): confluent cultures. Homogeneous Balb/c keratinocyte cultures are (a and b); Homogeneous C3H/HeN keratinocyte cultures are (c and d); 50%-50% chimeric cultures are (e and f) and 25%-75% chimeric cultures are (g and h). Note the cuboidal and regular morphology of the cells in the homogeneous as well as in the chimeric cultures. Magnification x 200.

using specific monoclonal antibodies in order to assess if the initial seeding ratios were still the same. Chimeric cell suspensions were first treated with an anti-H2^k monoclonal antibody (specific for C3H/HeN cells), and exposed to a goat anti-mouse IgM-IgG-fluorescein (FITC) conjugated in the dark. Second, cells were treated with an anti-H2^d monoclonal antibody (specific for Balb/c cells) coupled to phycoerytrin (PE). After these different treatments, cells were fixed in 1% paraformaldehyde solution and analyzed by flow cytometry. As controls, homogeneous Balb/c and C3H/HeN confluent keratinocyte cultures were enzymatically dissociated, then stained separately with an anti-H2^d-PE labeled or H2^k plus IgM-IgG-FITC conjugated antibodies, respectively. Percentages were plotted in Figure 5.3. These results showed that, at confluence, chimeric cultures contained the original proportions of each cell type. Indeed, from 50% Balb/c keratinocytes with 50% C3H/HeN keratinocytes seeded cultures (ratio: 1/1), we had 40% Balb/c keratinocytes with 42% C3H/HeN keratinocytes at confluent cultures (ratio: 1/1). Also, 25% Balb/c-75% C3H/HeN (ratio: 1/4) culture gave 18% Balb/c-60% C3H/HeN (ratio: 1/4). These immunostaining analyses confirmed the presence of both keratinocyte phenotypes in the confluent culture and demonstrate that there was no growth inhibition of the Balb/c keratinocytes when cocultured with the C3H/HeN keratinocytes and vice versa. Consequently, the chimeric epithelial substitutes ready for grafting were composed of both cell types in the same ratio as the initial cell plating. After these analyses, chimeric and homogeneous epidermal substitutes were produced and grafted.

SKIN REPLACEMENT USING CHIMERIC EPIDERMAL SUBSTITUTES

Graftable Epidermal Substitute Production

Graftable epidermal substitutes were prepared when keratinocyte cultures had reached confluence as described previously.[17,24] Confluent keratinocyte cultures were washed twice with PBS; then treated with Dispase (2.5 mg/ml) solution at 37°C for 20 to 30 min to release epidermal substitutes from the flask's surface. To facilitate handling these epidermal substitutes were transferred to a sterile Vaseline gauze basal side up and fixed together. These prepared epidermal substitutes were left immersed in the appropriate medium until grafting.

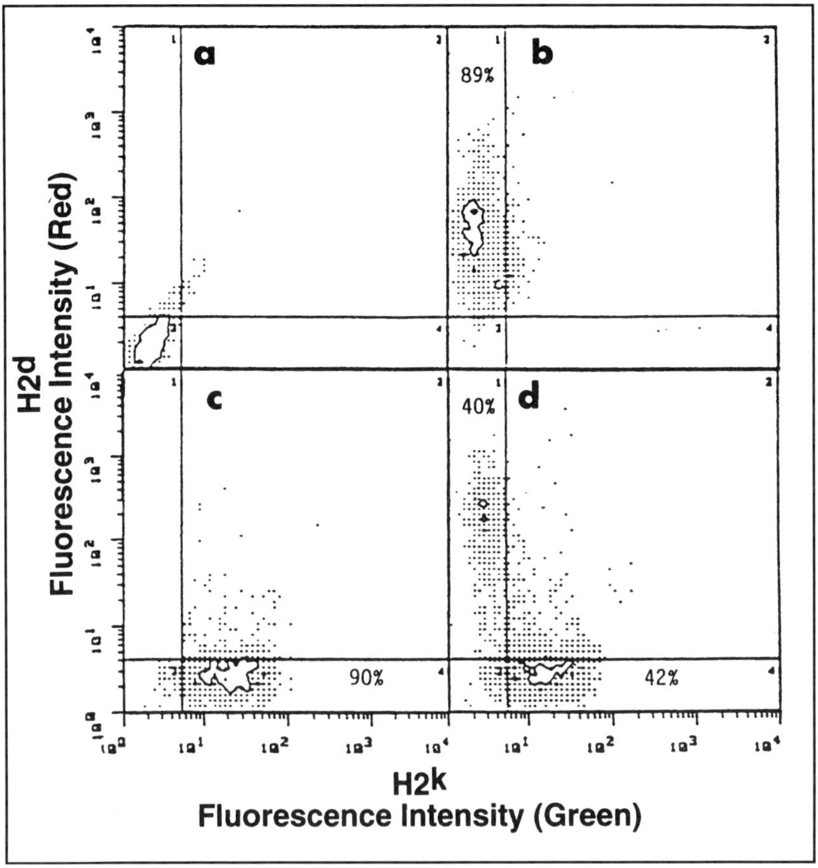

Fig. 5.3. Determination of the cell ratio in the confluent chimeric cultures. Confluent homogeneous (100% Balb/c or C3H/HeN) and chimeric (50%-50%; 75%-25% or 25%-75%) cultures were resuspended following enzymatic digestion. Cells were then stained separately or double stained with anti-H2^d (for Balb/c) and anti-H2^k (C3H/HeN) monoclonal antibodies, respectively, then analyzed using flow cytometry. (c, d) represent Chimeric (50%-50% and 25%-75%) cultures respectively. (a) represents the negative control (keratinocytes stained with an irrelevant primary antibody); (b) represents homogeneous cultures stained with anti-H2^d monoclonal antibody. The same results were obtained with homogeneous C3H/HeN keratinocyte cultures. Reprinted with permission from Rouabhia M et al. Transplantation 1995; 59(9):122-1235.

EPIDERMAL SUBSTITUTE GRAFTING

As shown in Figure 5.4, different epidermal substitutes were produced and grafted. For chimeric implant grafting, the recipient always refers to the smallest percentage of keratinocyte in the chimeric culture. Indeed, as pointed out in Table 5.2, for the 50% Balb/c-50% C3H/HeN implants, recipient mice were adult Balb/c or C3H/HeN. However, for 25% Balb/c-75% C3H/HeN implants, recipients are only adult Balb/c mice, and for 25% C3H/HeN-75% Balb/c implants, recipients are adult C3H/HeN mice.

Mice were anesthetized and each graft bed was prepared as described previously,[17,24] following the guidelines of the Canadian Council on Animal Care. To provide a physical barrier and prevent host re-epithelialization, a biocompatible sterile silicone chamber (2.5 cm diameter) was soaked in a saline antibiotic solution and then stabilized in situ by simple stitching, as previously described.[17,24,28] Homogeneous or chimeric epidermal substitute was deposited on the muscle (Fig. 5.4). The top of the transplantation chamber was installed to prevent the implant from drying. Five days later, all transplantation chamber tops were removed, allowing macroscopic assessments of the isografts, allografts and chimeric implants. In all experiments macroscopic observations showed that, up to 9-11 days after transplantation, foreign (100% allogeneic cells) and chimeric (50%-50%, 25%-75% or 75%-25%) epidermal substitutes survived as well as isografted epidermal substitutes (data not shown). No signs of necrotic zones were seen, as judged by visual inspection. On days 12 to 13 of grafting, all allografts showed necrotic zones. Rejection was heralded by a rapid change of the graft color from nice pink to deep purple, then brown, followed by the disintegration of the allografts. Finally, complete denudation of the grafting bed was observed. However, the chimeric implants showed the same macroscopic morphology (nice pink color, no necrotic or brown zones) as the isografts. The percentage of surface covered after chimeric epidermal substitute grafting confirms its healthy state. Thus, 14 days after grafting the percentage of permanently covered surface is more than 80%. This high coverage and skin replacement was confirmed 30 days post-grafting and was comparable to those obtained with isografted homogeneous epidermal substitute. These interesting results beg the question about the newly generated skin structure.

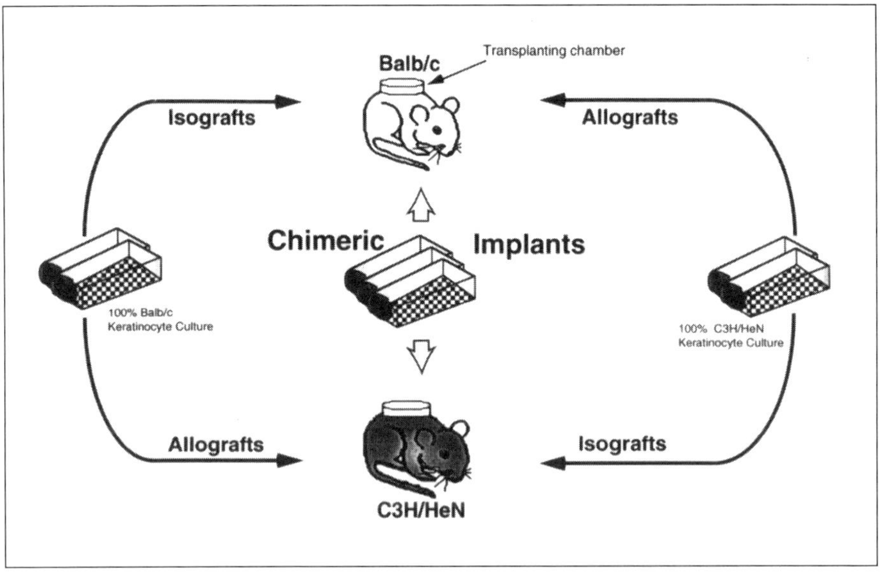

Fig. 5.4. Diagram summarizing the production and grafting of different epidermal substitutes.

Table 5.2. Different epidermal substitute production and grafting

Epidermal Substitutes	Final Percentage	Recipients	Graft Types
Homogeneous			
1	100%B	Balb/c	Isograft
2	100%C	Balb/c	Allograft
Chimeric			
1	50%B-50%C	Balb/c or C3H/HeN	Chimeric implant
2	25%B-75%C	Balb/c	Chimeric implant
3	75%B-25%C	C3H/HeN	Chimeric implant

B: Balb/c; C: C3H/HeN

Histological Aspect of the Newly Generated Skin After Chimeric Epidermal Substitute Grafting

Biopsies from the allografts, isografts and chimeric implants (50-50%, 25-75% or 75-25%) were processed on day 14. They were fixed and paraffin embedded. Thereafter, thin sections were stained with hematoxylin-eosin, observed under a Nikon Optiphot microscope and photographed. This histological study showed complete degradation of the allografts, but a well-organized epidermis after chimeric grafts (Fig. 5.5). This new cutaneous tissue showed the different known strata (corneum, granulosum, spinosum and germinativum). These histological features were confirmed at 30 days (Fig. 5.6). Epidermal chimeric structures are comparable to those obtained with the isografts. Hence, damaged skin may be successfully replaced using chimeric epidermal substitute. Nevertheless, the two relevant questions that emerged from this histological study are:

1. Do these newly generated cutaneous tissues, after chimeric epidermal substitute grafting, secrete skin proteins such as laminin

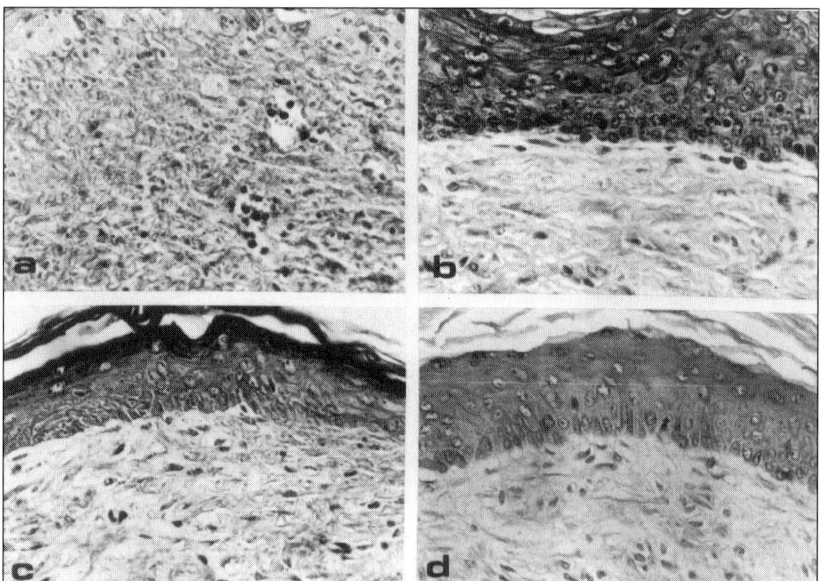

Fig. 5.5. Histological aspects of the regenerated epidermis after 14 days of chimeric implant grafting. Biopsies were harvested 14 days later for histological studies. (a) Allograft histology. (b) Isograft histology. (c) Chimeric (50-50%) implant histology. (d) Chimeric (25-75%) implant histology. The grafting of chimeric implants onto Balb/c recipient mice led to the same results. Magnification X250. Reprinted with permission from Rouabhia M et al. Transplantation 1995; 59(9):122-1235.

and type-IV collagen that are essential to the integrity of a healthy dermo-epidermal junction?

2. Are these newly generated cutaneous tissues well vascularized?

Protein Synthesis by the Newly Generated Skin

To answer the above questions, intact biopsies were harvested 14 days post-grafting from isografts and chimeric (50-50%, 25-75% or 75-25%) implants, embedded in OCT compound, then thin sections were prepared from each biopsy. They were stained with specific monoclonal antibodies to laminin and type IV collagen proteins, then examined using a fluorescence microscope and photographed. This staining showed that cells of the newly generated skin, after chimeric epidermal substitute grafting, synthesize laminin and type IV collagen proteins (Fig. 5.7). These protein secretions are comparable to those obtained with the isografts. Laminin and type IV collagen were present and also well secreted after 30 days of chimeric epidermal substitute grafting (Fig. 5.8). They confirm the well-structured

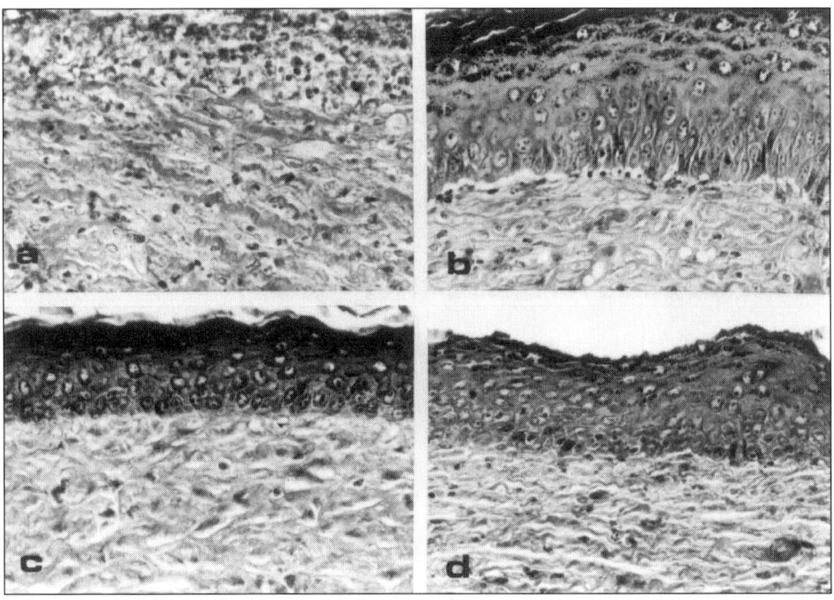

Fig. 5.6. Histological aspects of the regenerated epidermis after 30 days of chimeric implant grafting. Biopsies were harvested 30 days later for histological studies. (a) Allograft histology. (b) Isograft histology. (c) Chimeric (50-50%) implant histology. (d) Chimeric (25-75%) implant histology. The grafting of chimeric implants onto Balb/c recipient mice led to the same results. Magnification X250. Reprinted with permission from Rouabhia M et al. Transplantation 1995; 59(9):122-1235.

skin with a dermo-epidermal junction. To assess the degree of vascularization of the newly generated skin, biopsy samples were stained using an anti-CD31 monoclonal antibody. This antibody reacts specifically with a constitutively expressed protein on the surface of endothelial cells. As shown (Fig. 5.9), newly generated skin after chimeric epidermal substitute grafting was as well vascularized as the isograft generated tissues. The evaluation of vessel number per square millimeter in the chimeric and isograft biopsy samples confirmed (data not shown) the important degree of vascularization in both implant types with no significant differences between both grafts. The presence of laminin and type IV collagen proteins and the vascularization in the newly generated cutaneous tissue confirmed that the chimeric epidermal substitutes allow well structured and functional skin

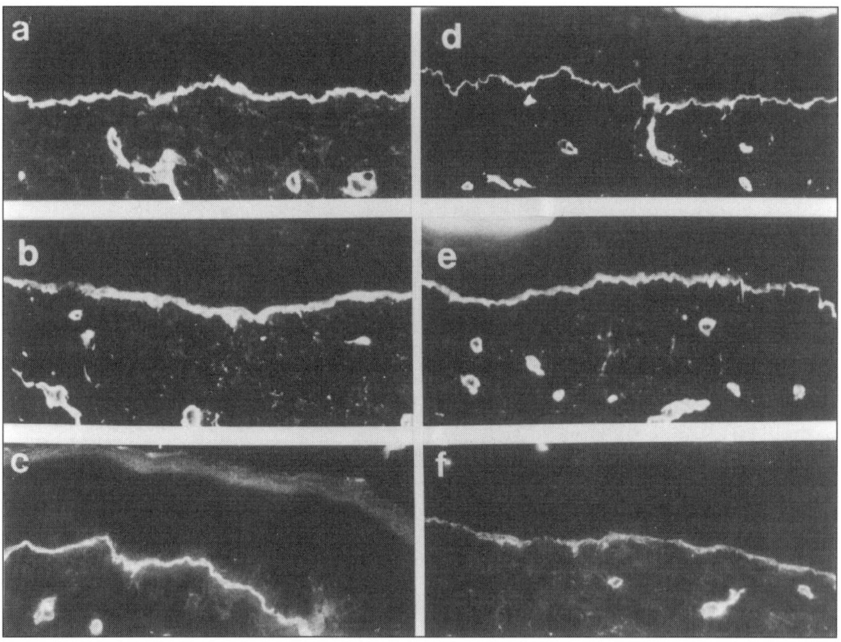

Fig. 5.7. Laminin and type IV collagen synthesis into generated skin 14 days after chimeric epidermal substitute grafting. Biopsies were taken 14 days postgrafting and stained with anti-laminin or anti-type IV collagen specific antibodies. Note the linear deposition of immunoreactants along the dermo-epidermal junction for laminin (a,b,c) and type IV collagen (d,e,f). (b,e) Chimeric (50-50%) and (c,f) chimeric (25-75%) implants. These stainings were comparable to those obtained with the isografts (a,d). Magnification X200. Reprinted with permission from Rouabhia M et al. Transplantation 1995; 59(9):122-1235.

replacement. The question is: Does the newly generated tissue (epidermis) contain both allogeneic and syngeneic keratinocytes?

KERATINOCYTE PHENOTYPING

Intact biopsies were harvested 30 days postgrafting from isografts and chimeric (50-50%, 25-75% or 75-25%) implants, embedded in OCT compound and stored frozen. Thin sections were prepared from each biopsy and stained with MHC class I specific monoclonal antibodies for Balb/c keratinocytes (anti-H2^{d}) and C3H/HeN keratinocytes (anti-H2^{k}). As presented in Figure 5.10, the regenerated epidermis from C3H/HeN mice grafted with chimeric epidermal substitute showed major and uniform staining (syngeneic keratinocytes), such as the isografts. However, very low and reduced unstained surfaces

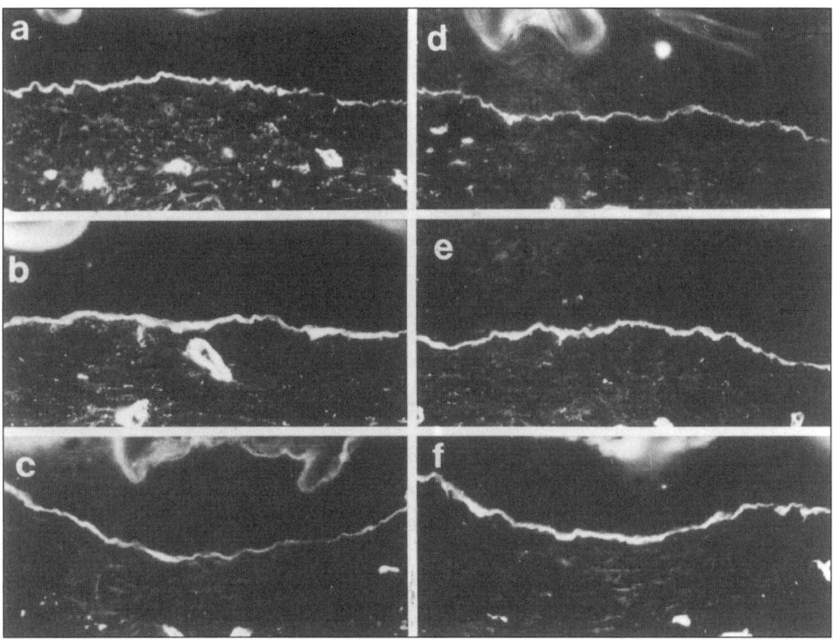

Fig. 5.8. Laminin and type IV collagen synthesis into generated skin 30 days after chimeric epidermal substitute grafting. Biopsies were taken 30 days postgrafting and stained with anti-laminin or anti-type IV collagen specific antibodies. Note the linear deposition of immunoreactants along the dermo-epidermal junction for laminin (a, b, c) and type IV collagen (d, e, f). (b, e) Chimeric (50-50%) and (c, f) chimeric (25-75%) implants. These stainings were comparable to those obtained with the isografts (a, d). Magnification X200. Reprinted with permission from Rouabhia M et al. Transplantation 1995; 59(9):122-1235.

were present in some experiments especially after 14 days postgrafting (data not shown), suggesting the presence of allogeneic keratinocytes in the newly regenerated skin. Two hypotheses emerge from these results:

1. Allogeneic keratinocytes present in the chimeric implants serve as a temporary vehicle for autologous keratinocyte proliferation; then these epidermal cells disappear with the normal shedding mechanism, since the epidermis is a constantly renewing tissue. Indeed, allogeneic keratinocytes may stimulate autologous

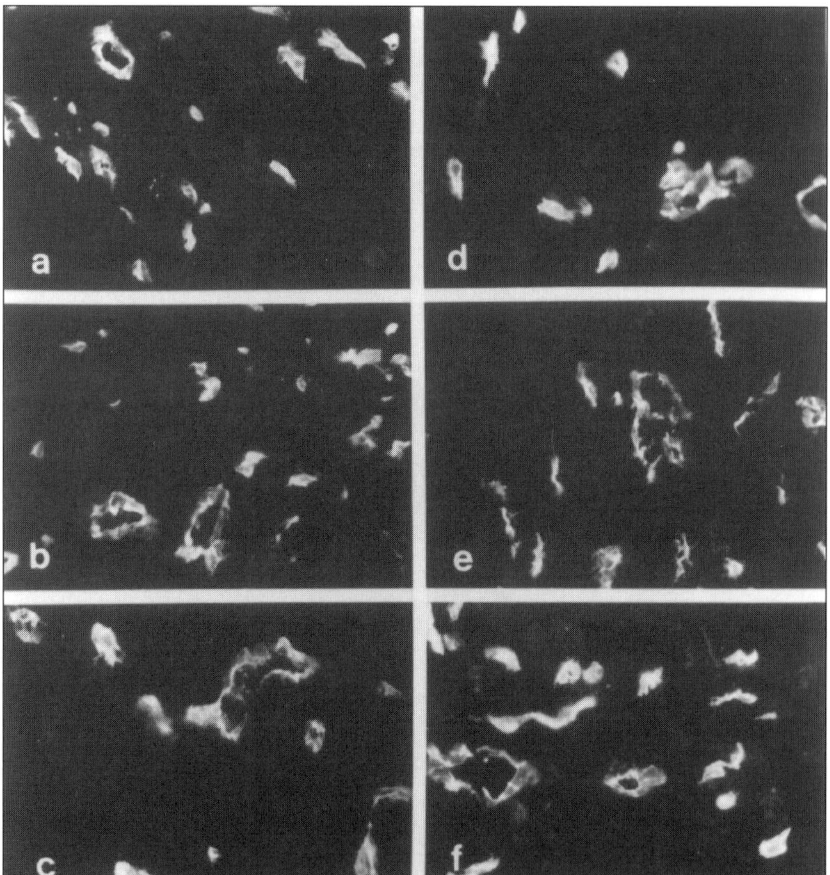

Fig. 5.9. Tissue vascularization. Biopsy specimens were taken 14 (a,b,c) and 30 (d,e,f) days after grafting and stained with anti-CD31 monoclonal antibody. This staining showed immunoreactant deposition around vessels in the newly generated tissues after isografts (a,d), 50%-50% chimeric (b,e) and 25%-75% chimeric (c,f) implant grafting. Magnification: X300.

epidermal cells to proliferate through (i) a cell-cell contact, and via (ii) cytokine release. However, we still have to explore the function(s) of allogeneic keratinocytes present in the chimeric epidermal implants before their normal shedding.

2. Allogeneic keratinocytes present in the chimeric epidermal substitutes were immunologically eliminated from the graft via selective mechanisms. Clinical results obtained after cultured epidermal allograft onto second degree burn patients strongly suggest that allogeneic keratinocytes were progressively and insidiously replaced by the recipient keratinocyte ingrowth without any evidence of allogeneic keratinocyte rejection.[29-31] In this clinical setting the residual appendages provide a steady stream of autologous keratinocytes that will creep out and replace the overlaying allogeneic cells. Thus, after a few weeks all allogeneic cells were eliminated and no engraftment was noted. Identical clinical results were obtained by covering donor sites with allogeneic cultured epidermal substitutes. These two therapeutic results, suggest that when cultured epidermal allografts are implanted onto deep second degree burns, where deep portions of hair follicles are still intact and contain autologous keratinocytes,

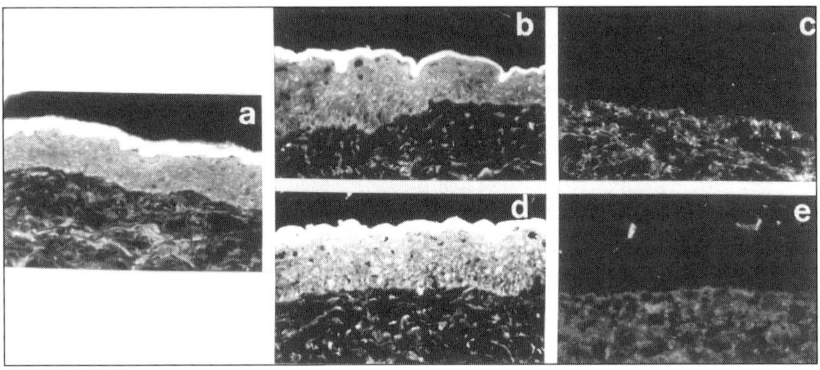

Fig. 5.10. Phenotype of the epithelial cells in the newly generated epidermis after chimeric implant grafting. Biopsies were taken from the isografts and the chimeric (50%-50%; 75% -25% or 25%-75%) implants after 30 days postgrafting onto C3H/HeN mice then stained with anti-H2^d or anti-H2^k monoclonal antibodies. (a) isograft, (b,c) 50%-50% chimeric implants, (d,e) 75%-25% chimeric implants. Note that biopsies taken from chimeric implants and labeled with H2^k (b,d) were specifically positive as the isograft biopsies (a). However, their labeling with anti-H2^d (c,e) was specifically negative. Magnification: X200. Reprinted with permission from Rouabhia M et al. Transplantation 1995; 59(9):122-1235.

autologous epithelial cells proliferate beneath the allograft and cover the wound surface. These clinical studies and other experimental results strengthen our second hypothesis. Consequently, immunological study is necessary.

IMMUNOLOGICAL STUDIES AFTER CHIMERIC EPIDERMAL SUBSTITUTE GRAFTING

The involvement of lymphoid cells in the elimination of histoincompatible cells has been established by adoptive transfer and diffusion chamber studies.[32-36] Infiltration of the graft parenchyma by mononuclear cells, producing a destructive lesion, is common in allograft and allograft rejection as well as in delayed-type hypersensitivity. Depending on the histological type of the transplant, its anatomical site, and the time of examination, the infiltrating cells could be observed in varying numbers. Because the implants contained syngeneic and allogeneic cells, two questions emerge from these studies: how does leukocyte infiltration in the chimeric implants compare with that observed in allografts? What is the phenotype of these leukocytes in the infiltrate?

Cellular Immunity

To answer the first question, hematoxylin and eosin-stained chimeric biopsy specimens showed (Fig. 5.11) an important leukocyte infiltration of the allografts, indicating their rejection. However, leukocyte infiltration was present in weak numbers in the chimeric implants 14 days after grafting allowing well-structured skin replacement. Thirty days after grafting, the leukocytes infiltrating the chimeric implant were still present (Fig. 5.11), but seemed to be reduced in number. To answer the second question, immunofluorescence analyses were done to phenotypically identify the white cells infiltrating the implants. As shown in Table 5.3, this infiltrate contained a large number of monocytes (Mac-1^+) in the allografts, while a weak monocyte infiltration was noted in the chimeric implants. Using specific T cell antibodies, immunofluorescence staining revealed that the leukocyte infiltrate contained $CD4^+$ and $CD8^+$ T cells in the allograft as well as in the chimeric implants. The presence of infiltrating leukocytes was also found in the chimeric implant 30 days after grafting (Table 5.4); their number seems to be lower than in chimeric implants 14 days after grafting. The presence of leukocyte infiltration in the chimeric implant suggests cellular immunity activation. This cellular involvement could be followed by humoral immunity activation.

Humoral Immunity

Graft rejection is initiated through cellular immunity as well as through humoral immunity via antibody dependent cytotoxicity.[37-40] Consequently, signs of humoral immunity in allogeneic cell rejection

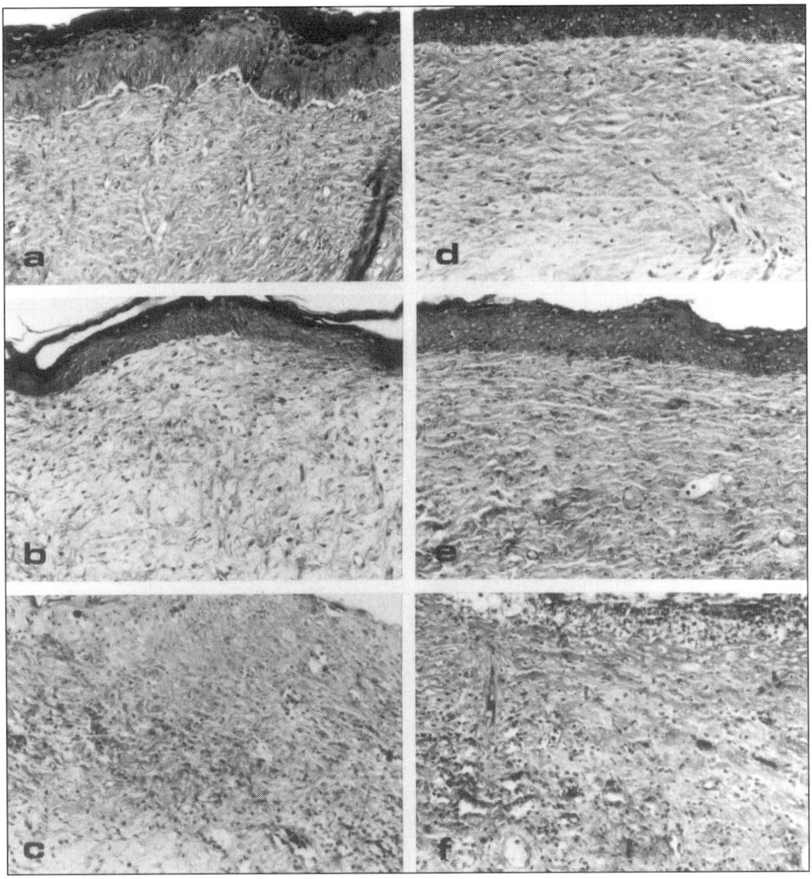

Fig. 5.11. Leukocyte infiltration in the newly generated cutaneous tissue after chimeric epidermal substitute grafting. Biopsies were performed 14 (a,b,c) and 30 (d,e,f) days after epidermal substitute grafting for infiltrating leukocyte assessment using hematoxylin and eosin staining. a, d: Infiltrating cells in 50%-50% chimeric implants. (b,e) Infiltrating cells in 25%-75% chimeric implants. (c,f) Infiltrating cells in the allografts. Magnification: X150. Reprinted with permission from Rouabhia M et al. Transplantation 1995; 59(9):122-1235.

Table 5.3. Phenotypic features of infiltrating leukocytes 14 days postgrafting

Implants	Mac-1	CD8	CD4
Isografts	–	–	–
Allografts	++++	++	++
Chimeric 50%-50%	++	+	+
Chimeric 25%-75%	++	+	+

Table 5.4. Phenotypic features of infiltrating leukocytes 30 days postgrafting

Implants	Mac-1	CD8	CD4
Isografts	–	–	–
Allografts	+++	++	++
Chimeric 50%-50%	++	+/–	+/–
Chimeric 25%-75%	++	+/–	+/–

may occur after chimeric sheet grafting. To clarify this situation, recipient sera were used to stain human normal skin biopsy specimens. Presented in Figure 5.12, the results reveal the presence of antibodies reacting with the human keratinocytes (epidermis). This positive staining suggests the presence of circulating antibodies reacting specifically with allogeneic keratinocytes. This response, coupled with cellular immunity, results in the selective elimination of all allogeneic keratinocytes present in the chimeric implants.

CLINICAL IMPACT OF THIS CHIMERIC CULTURE METHOD

Our own data and those in the literature show that autologous cultured epidermal cells and grafting are the only life-saving therapy for the patient with massive burns. But keratinocyte growth on a large scale needs several weeks. During this critical time, the patient is suffering, the risk of infection is high and the physician, nurses and family are under continuous stress to keep the burn patient alive until his own cultured epidermal substitutes are ready. Consequently, reducing the autologous cultured time would be a major improvement. Thus, clinical application of chimeric (mixture of allogeneic and

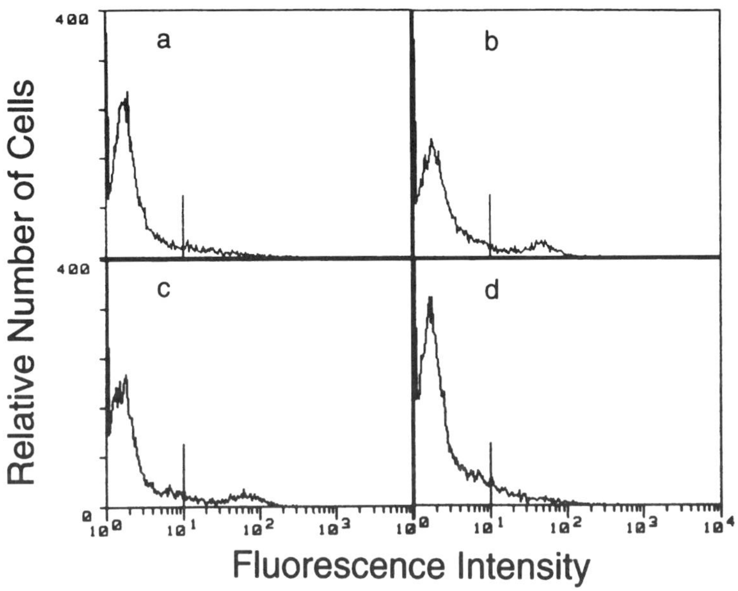

Fig. 5.12. Specific staining of donor keratinocytes with chimeric recipient sera. Sera were collected from recipient mice after 14 days of chimeric (50%-50%) and (25%-75%) implants and used to stain donor keratinocytes. Donor cells refers to Balb/c when chimeric epidermal substitutes were grafted onto C3H/HeN mice and vice-versa. Control sera were also collected from isografts and allografts then used.

autologous keratinocyte populations) epidermal substitutes for large burn wound treatment could be a significant therapeutic advance. The use of this new approach for burned patient treatment can decrease by at least half the previously described delay for epidermal culture and significantly reduce the cost of burn management during patient hospitalization. Compared to the standard method (autologous cultured sheets), chimeric epithelia can be readily produced by using primary subcultures of autologous keratinocytes at lower (<50%) confluence percentage. Banks of human allogeneic keratinocytes, previously tested for virus contamination and other infectious micro-organisms, could be established and available at all times (Fig. 5.13). Autologous and allogeneic cells could be mixed at an adequate ratio, plated in flasks and incubated for graftable sheet production. Graftable epidermal substitutes would be obtained after a short time (<2 weeks), then the patient would be treated sooner

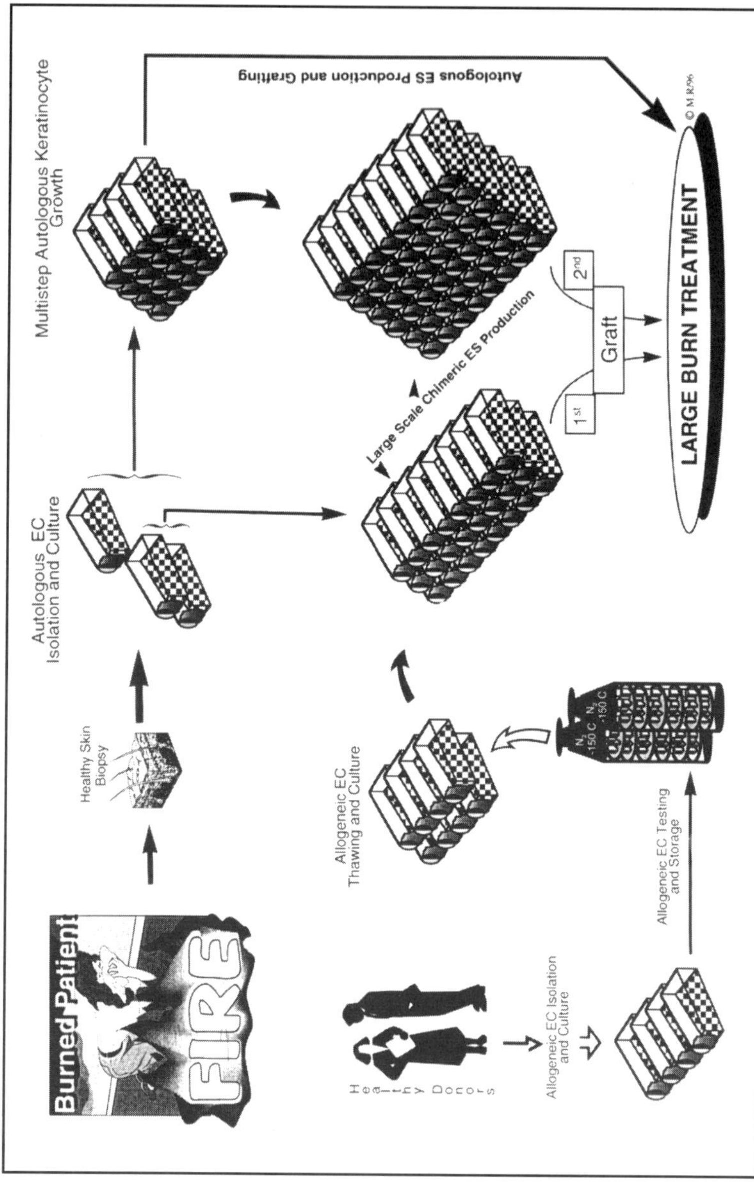

Fig. 5.13. Potential application of chimeric epidermal cell culture for burn treatment.

with complete (autologous-allogeneic) coverage. Before their elimination, allogeneic keratinocytes probably secrete different stimulating growth factors. These proteins will stimulate the patient's keratinocytes to proliferate and passively replace allogeneic keratinocytes as shown in the second degree burn or donor site grafted with complete allogeneic epidermal substitutes. This cell culture method is a real innovation and a major improvement for damaged skin replacement and also for several fundamental studies in organ and tissue transplantation.

FUNDAMENTAL IMPACT OF THIS CHIMERIC CULTURE METHOD

This new chimeric culture technology could find multiple applications in fundamental assessment of organ and tissue transplantation. Since the ultimate aim in allogeneic transplantation is to achieve an immunological tolerance[41-46] between the donor and recipient, our chimeric culture approach may provide ways to study tolerance phenomena on cutaneous tissue. Interesting clinical achievements have been obtained with allogeneic liver transplantation suggesting that this organ itself may produce tolerogenic factors that may reduce or eliminate the need for patient immunosuppression.[47,48] The presence of allogeneic cells in the chimeric implants may serve as a temporary vehicle for autologous keratinocytes proliferation by a cell-cell contact or through cytokine release.[49] It seems likely that the allogeneic keratinocytes will disappear with the normal shedding mechanism, since the epidermis is a constantly renewing tissue. This hypothesis was supported by our studies, herein described, where chimeric implants were grafted and showed cutaneous structure regeneration without rejection signs. However, much needs to be learned about the chimeric donor cell function before their passive elimination, the different mechanisms involved in this passive elimination and the meaning of tolerance and the clonal silencing after skin allotransplantation. This chimeric epithelial implant may help us to better understand some of the mechanisms involved in skin allograft tolerance.

CONCLUSION

The chimeric culture method demonstrates that graftable chimeric (allogeneic-syngeneic) substitutes allow skin replacement.

The newly generated cutaneous tissue was well structured and physiologically active. Although the recipient's cellular and humoral immunity were activated, skin was nicely replaced with no rejection signs of allogeneic cells initially present in the chimeric epidermal substitutes. Do chimeric (syngeneic-xenogeneic) epidermal substitutes allow skin replacement as did syngeneic-allogeneic implants? The answer to this question is in chapter 6.

Acknowledgments

The author thanks Prof. S. William A. Gunn for critically reviewing this chapter. This research was supported financially by the Medical Research Council of Canada and the Fonds de la Recherche en Santé du Québec.

References

1. Tanner JC, Vandeput J, Olley JF. The mesh skin graft. Plast Rec Surg 1964; 34:287-292.
2. Phillips TJ, Kehinde O, Green H et al. Treatment of skin ulcers with cultured epidermal allografts. J Am Acad Dermatol 1989; 21:191-199.
3. Alexander JW, MacMillan BG, Law EJ et al. Treatment of severe burns with widely meshed skin autograft and meshed skin allograft overlay. J Trauma 1981; 21:433-438.
4. Heimbach DM. Early burn wound excision and grafting. In: Boswick J, ed. The Art and Science of Burn Care. Rockville, Md: Aspen Publishers, 1987:65-73.
5. Dagum AB, Peters WJ, Neligan PC et al. Severe multiple mononeuropathy in patients with major thermal burns. J Burn Care Rehabil 1993; 14(4):440-445.
6. Greenleaf G, Cooper ML, Hansbrough JF. Microbial contamination in allografted wound beds in patients with burns. J Burn Care Rehabil 1991; 12(5):442-445.
7. Freeman AE, Igel HJ, Waldman NL. A new method for covering large surface area wounds with autografts. Arch Surg 1974; 108:721-723.
8. Fang C-H, Alexander JW. Wound contraction following transplantation of microskin autografts with overlaid skin allograft in experimental animals Burns 1990; 16(3):190-192.
9. Winkelstein A. What are the immunological alterations induced by burn injury? J Trauma 1984; 24:72s-83s.
10. Gallico GG, O'Connor NE, Compton CC et al. Permanent coverage of large burn wound with autologous cultured human epithelium. N Engl J Med 1984; 311:448-451.
11. Green H, Kehinde O, Thomas J. Growth of cultured human epidermal cells into multiple epithelia suitable for grafting. Proc Natl Acad Sci USA 1979; 76:5665-5668.

12. Shirani KZ, Vaughan GM, Mason Jr AD et al. Update on current therapeutic approaches in burns. Shock 1996; 5(1):4-16.
13. Rue LW III, Cioffi Jr WG, McManus WF et al. Wound closure and outcome in extensively burned patients treated with cultured autologous keratinocytes. J Trauma 1993; 34:662-668.
14. Hefton JM, Madden MR, Finkelstein JL et al. Grafting of burn patients with allografts of cultured epidermis cells. Lancet 1983; 2:428-430.
15. Thivolet J, Faure M, Demidem A et al. Long-term survival and immunological tolerance of human epidermal allografts produced in culture. Transplantation 1986; 42:274-280.
16. Faure M, Mauduit G, Schmitt D et al. Growth and differentiation of human epidermal cultures used as auto and allografts in humans. Br J Dermatol 1987; 116:161-170.
17. Rouabhia M, Germain L, Bélanger F et al. Cultured epithelium allografts: Langerhans cell and Thy-1^+ dendritic epidermal cell depletion effects on allograft rejection. Transplantation 1993; 56(2):259-264.
18. Auböck J, Irschick E, Romani N et al. Rejection after a slightly prolonged survival time, of Langerhans cell-depleted allogeneic cultured epidermis used for wound coverage in humans. Transplantation 1988; 45:730-737.
19. Phillips TJ. Cultured epidermal allografts—a permanent or temporary solution? Transplantation 1991; 51:937-941.
20. Hultman CS, Cairns BA, deSerres S et al. Burn Injury impairs second-set rejection and CTL reactivity in mice primed by cultured keratinocyte allografts. Transplantation 1995; 60(6):584-589.
21. Gaspar AA, Katz SI. Induction and functional characterization of class II MHC (Ia) antigens on murine keratinocytes. J Immunol 1988; 140:2956-2963.
22. Bos JD, Kapsenberg ML. The skin immune system: progress in cutaneous biology. Immunol Today 1993; 14(2):75-78.
23. Green H, Kehinde O, Thomas J. Growth of cultured human epidermal cells into multiple epithelia suitable for grafting. Proc Natl Acad Sci USA 1979; 76:5665-5668.
24. Rouabhia M, Germain L, Bélanger F et al. Optimization of murine keratinocyte culture for the production of graftable epidermal sheets. J Dermatol 1992; 19(6):325-334.
25. Rheinwald JG, Green H. Serial cultivation of strains of human epidermal keratinocytes: the formation of keratinizing colonies from single cells. Cell 1975; 6:331-337.
26. Holbrook KA, Hennings H. Phenotypic expression of epidermal cells in vitro: a review. J Invest Dermatol 1983; 81:11s-24s.
27. Boyce ST, Ham RG. Cultivation, frozen storage, and clonal growth of normal human epidermal keratinocytes in serum-free media. J Tissue Cult Meth 1985; 9:83-93.

28. Xu W, Germain L, Goulet F et al. Permanent grafting of living skin substitutes: surgical parameters to control for successful results. J Burn Care Rehabil 1996; 17:7-13.
29. Phillips TJ, Bhawan J, Leigh IM et al. Cultured epidermal autografts and allografts: A study of differentiation and allograft survival. J Am Acad Dermatol 1990; 23:189-198.
30. Leigh IM, Purkis PE, Navsaria H et al. Treatment of chronic venous ulcers with sheets of cultured allogeneic keratinocytes. Br J Dermatol 1987; 117:591-597.
31. Phillips TJ, Gilchrest BA. Cultured allogeneic keratinocyte grafts in the management of wound healing: prognostic factors. J Dermatol Surg Oncol 1989; 15:1169-1176.
32. Mason DW, Dallman MJ, Arthur RP et al. Mechanisms of allograft rejection: the roles of cytotoxic T-cells and delayed-type hyper-sensitivity. Immunol Rev 1984; 77:167-184.
33. Auchincloss Jr H, Sultan H. Antigen processing and presentation in transplantation. Curr Opin Immunol 1996; 8(5):681-687.
34. Hayry P, von Willebrand E, Parthenais E et al. The inflammatory mechanisms of allograft rejection. Immunol Rev 1984; 77:85-142.
35. Chandler C, Passaro E Jr. Transplant rejection: Mechanisms and treatment. Arch Surg 1993; 128:279-283.
36. Tilney NL, Strom TB, Macpherson SG et al. Surface properties and functional characteristics of infiltrating cells harvested from acutely rejecting cardiac allografts in inbred rats. Transplantation 1975; 20:323-330.
37. Winn HJ, Baldamus CA, Jooste SV et al. Acute destruction by humoral antibody of rat skin grafted to mice: The role of complement and polymorphonuclear leukocytes. J Exp Med 1973; 137:893-910.
38. Whitley D, Kupiec-Weglinski W, Tileny NL. Antibody-mediated rejection of organ grafts. Curr Opin Immunol 1990; 2:864-869.
39. Jooste SV, Colvin RB, Winn HJ. The vascular bed as the primary target in the destruction of skin grafts by antiserum. II. Loss of sensitivity to antiserum in long-term xenografts of skin. J Exp Med. 1981; 154:1332-1341.
40. Manning DD, Reed ND, Shaffer CF. Maintenance of skin xenograft of widely divergent phylogenetic origin on congenitally athymic (nude) mice. J Exp Med 1973; 138:488-494.
41. Billingham RE, Brent L, Medawar PB. Actively acquired tolerance to foreign cells. Nature 1953; 172:603-606.
42. Houssaint E, Flajnik M. The role of thymic epithelium in the acquisition of tolerance. Immunol Today 1990; 11:357-360.
43. Shoskes DA, Wood KJ. Indirect presentation of MHC antigens in transplantation. Immunol Today 1994; 15:32-38.
44. Wood KJ. The induction of tolerance to alloantigens using MHC class I molecules. Cur Oppi Immunol 1993; 5:759-762.

45. Morris PJ, Dallman MJ, Wood KJ. Anergy in allogeneic transplantation. Transplant Proc 1993; 25:281-283.
46. Qin S, Cobbold SP, Pope H et al. "Infectious" transplantation tolerance. Science 1993; 259: 974-977.
47. Starzl ET, Demetris JA, Murase N et al. Donor cell chimerism permitted by immunosuppressive drugs: a new view of organ transplantation. Immunol Today 1993; 14:326-332.
48. Starzl ET, Demetris JA, Murase N et al. Cell migration, chimerism, and graft acceptance. Lancet 1992; 339:1579-1582.
49. Kratz G, Haegerstrand A, Dalsgaard CJ. Conditioned medium from cultured human keratinocytes has growth stimulatory properties on different human cell types. J Invest Dermatol 1991; 97:1039-1043.

CHAPTER 6

Damaged Skin Replacement Using Tissue Engineering: Chimeric (Syngeneic-Xenogeneic) Epidermal Substitute Production and Grafting

Mahmoud Rouabhia

INTRODUCTION

Over the past two decades, major advances in clinical application have been made in the field of transplantation. These advances are made possible through basic research in which alloreactive mechanisms are elucidated, allowing focused strategies to control the rejection response. The success of clinical allotransplantation has, however, revealed an entirely new problem; the shortage of organs' half life in vivo. Currently, the demand for organs and tissues greatly exceeds supply, justifying the great interest in the potential clinical application of xenotransplantation.

This interest stems from (i) the urgent need for donor organs and a sense that clinicians as well as the public are willing to consider xenotransplantation as a rational approach to the shortage of organs and tissues, and (ii) the progress in understanding the cellular and molecular events that contribute to xenograft rejection.

The concept of transplanting nonhuman organs or tissues into and onto human recipients is not new. Starting with Reemtsma's work,[1] transplantation of many different organs has become routine

Skin Substitute Production by Tissue Engineering: Clinical and Fundamental Applications, edited by Mahmoud Rouabhia.

from a technical point of view, although two major unresolved obstacles remain: xenograft rejection and the inadequate supply of organ and tissue donors.

XENOGRAFT REJECTION

The transplantation of organs or tissues from species into and onto other species was either immediately rejected or xenograft rejection occurred delayed.[2,3] Hyperacute rejection of nonmatched donor species is initiated by naturally occurring host antibodies targeted against glycoproteins expressed on the surface of the donor vascular endothelial cells.[4-6] However, the humoral response to xenoantigens is clearly not the only factor mediating xenorejection. In species combinations in which no preformed antibodies are present in the serum, rejection still occurs, although not quite as rapidly as in hyperacute rejection. In the absence of hyperacute rejection, vigorous cell-mediated graft rejection occurs in xenogeneic transplantation.[7-9] This rejection is still not controlled with immunosuppressive agents as it is in the allografts. Thus, the increasing enthusiasm about the prospects for clinical xenotransplantation is aimed at overcoming the xenorejection and increasing the adequate supply of donated organs and tissues.

SKIN XENOGRAFT: A CLINICAL NECESSITY

For the medical community, full- and deep partial-thickness skin defects present serious problems, with the patient highly vulnerable. Injuries with loss of skin, as occur often with burns, require special surgical strategies. It is necessary to consider the basis of the defect, necrectomy procedures, wound bed-preparation for grafting and the correct choice of temporary dressing.

The temporary cover of burns by biological dressing is presently the major improvement keeping patients alive and preparing the grafting bed for permanent skin replacement.[10,11] Mostly, human (cadaver skin) or animal (porcine skin) donors are solicited. However, the availability of cadaver skin is a limiting factor for burn dressing.[12] Pigskin is the most suitable for xenotransplantation having the closest resemblance to human skin in terms of thickness of the dermis, density, vascularization, composition and subcutaneous fat.[13,14] It is easy to handle since it can be transplanted and folded. This dressing was used by several groups to temporarily cover deep second and third degree burned patients.[14-16] It presents several advantages:

- Early application of skin xenografts prevents further superficial damage due to wound surface dehydration.
- Skin xenografts promote epithelialization at the wound edges and simultaneously suppress overgrowth of granulations in the center.
- The frequent dressing changes facilitate the debridement of the wound bed and prepare it for permanent closure.
- Due to their analgesic effect, xenografts are also valuable in the therapy of minor burns.

However, skin xenografts are limited due to high immunogenicity and rapid rejection.[17-19] Scientists working in the field of cutaneous transplantation are still faced with the challenge of at least trying to control the xenograft rejection as they do allograft rejection,[20,21] even if the mechanisms involved in the allotransplantation are not completely elucidated. For that reason, the chimeric culture approach may provide ways to shed light on some of the mechanisms involved in skin xenograft rejection, and may answer the question if it is possible to generate functional skin by grafting chimeric (syngeneic-xenogeneic) epidermal substitutes.

CHIMERIC (SYNGENEIC-XENOGENEIC) EPIDERMAL SUBSTITUTE PRODUCTION

Epidermal Cell Isolation

Murine epidermal cell isolation

Primary cultures of murine epidermal keratinocytes were established from neonatal Balb/c mouse skin as previously described.[22,23] Briefly, epidermis was separated from dermis after flotation of the skin on 0.25%-20 μg/ml of trypsin-DNase solution overnight at 4°C. The detached epidermal pieces were aseptically transferred to the medium containing serum to inhibit the action of residual enzyme and to mechanically release epidermal cells. Cell suspensions were applied onto lympholyte-M gradient and spun down to eliminate keratinous material. Live cells were counted and adjusted to the desired concentration.

Human epidermal cell isolation

Human epidermal cells were isolated from normal adult skin specimens[24] removed during reductive breast surgery of healthy subjects (20 to 40 years old). Biopsies were cut into small pieces and then digested in thermolysin solution overnight at 4°C. This treatment allows complete separation of the epidermis from the dermis. Epidermis was cut into small pieces and then incubated with 0.05% trypsin-0.1% EDTA for 30 min at 37°C. Human epidermal cells were washed, counted and resuspended in culture medium at the desired concentration. Before their use for chimeric epidermal substitute production, human keratinocytes were maintained in culture for at least two subcultures. They were used between the second and fourth passage to produce chimeric (syngeneic-xenogeneic) epidermal substitutes.

Chimeric Culture Performance

Different cultures were performed using human and murine keratinocytes separately (homogeneous cultures), or human and murine keratinocytes mixed together at different ratios:

1. 50% of human keratinocytes with 50% of murine keratinocytes
2. 75% human keratinocytes with 25% murine keratinocytes.

Cells were allowed to attach and proliferate in a 95% air, 5% CO_2 humidified atmosphere at 37°C. Cultures were observed and photographed daily, showing normal keratinocyte growth in the chimeric cultures as the keratinocyte growth in all homogeneous cultures (Fig. 6.1). Thus, murine keratinocytes proliferated and formed clones which were surrounded by human keratinocytes as shown at confluent stage (Fig. 6.1). These studies suggest the absence of growth dominance of human keratinocytes on murine keratinocytes and vice versa. The overgrowth phenomenon was investigated using the indirect immunofluorescence staining method.

Is There a Cell Growth Predominance?

Confluent culture was dissociated separately, and cell suspensions were stained using specific monoclonal antibodies to determine whether the initial seeding ratios were still the same after culture. Chimeric cell suspensions were double stained with anti-HLA-ABC polyclonal antibody (specific for human keratinocytes) anti-$H2^d$ or $H2^k$ monoclonal antibody (specific for Balb/c or C3H/HeN keratinocytes). Positive and negative cells were screened under a fluorescence

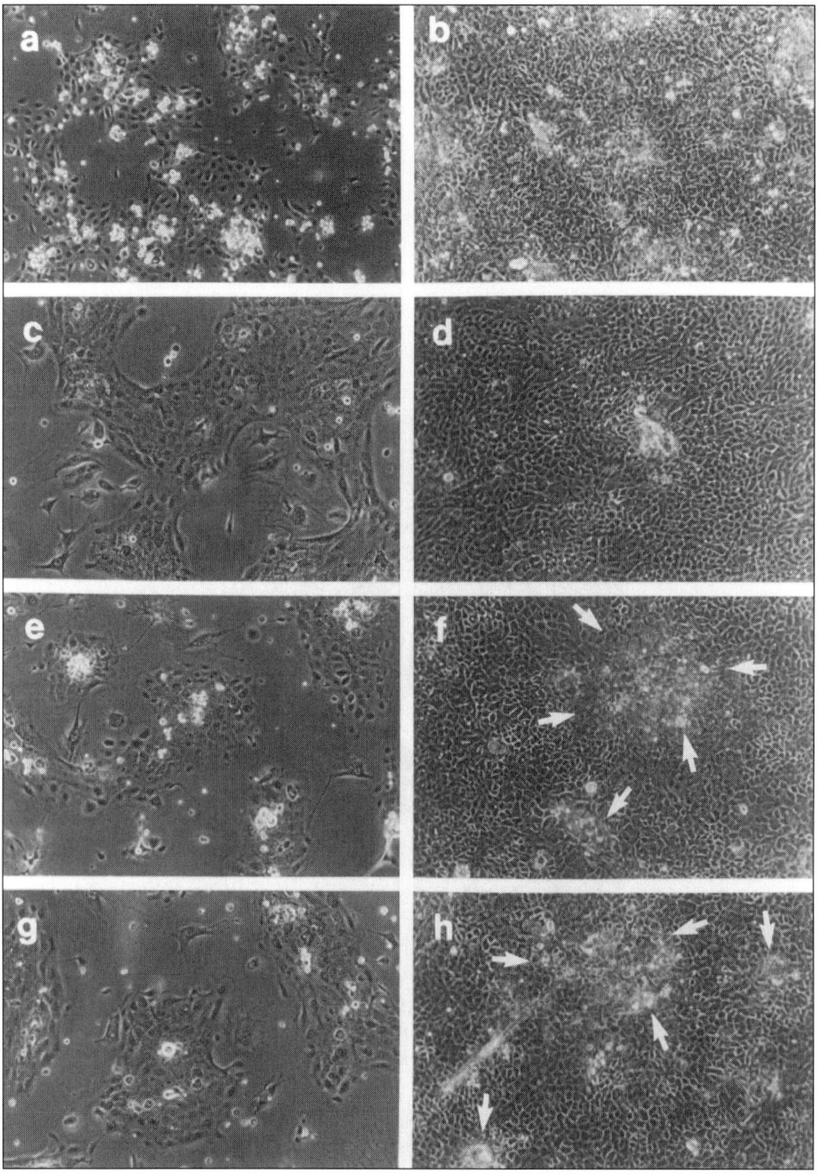

Fig. 6.1. Cell morphology in chimeric cultures. Epidermal cells were isolated from human and mouse (Balb/c or C3H/HeN) skin and seeded separately or cocultured at different ratios: 50%h-50%B and 25%B-75%h. a,b: 24 hours after seeding; c,d: at confluent stage. Homogeneous cultures are (a) and (c); chimeric culture are (b) and (d). Note the cuboidal and regular morphology of the cells in the homogeneous as well as in the chimeric cultures. Magnification x200.

microscope and plotted (Fig. 6.2). Thus, this staining confirms the presence of both cell types in the confluent cultures prior to grafting. Moreover, ratios of HLA-ABC labeled cells over $H2^d$ labeled cell ratios were well preserved. Indeed, from the initial 50%Balb/c-50%h and 25%Balb/c-75%h chimeric cultures, the percentages of each keratinocyte type prior to transplantation were 40%Balb/c-42%h and 17%Balb/c-60%h, respectively. The homogeneous (100%) human and murine keratinocyte cultures showed 87% and 85% labeled cells, respectively. This immunostaining did not show any growth inhibition of the Balb/c keratinocytes when cocultured with human keratinocytes and vice versa. Also, the chimeric epithelial sheets that were ready for

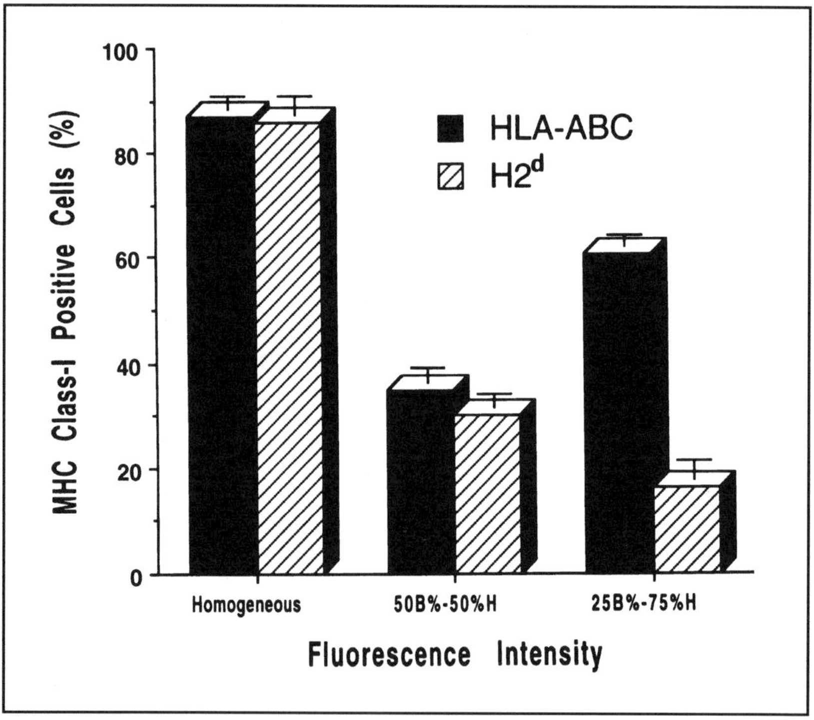

Fig. 6.2. Identification of human and murine keratinocyte percentages in confluent chimeric cultures prior to grafting. Cultured human, murine and chimeric (50%B-50%h; 25%B-75%h) cell suspensions were stained separately or double stained with Balb/c (anti-$H2^d$) and human (anti-HLA-ABC) monoclonal and polyclonal antibodies, respectively. The percentage (mean ± SD) of each cell type was plotted. Homogeneous histograms represent the percentage of $H2^d$ and HLA-ABC positive cells in the 100%Balb/c and human keratinocyte cultures, respectively. Reprinted with permission from Rouabhia M, Transplantation 1996; 61(9):1290-1300.

grafting were composed of both syngeneic and xenogeneic keratinocytes in the same ratio as the initial cell plating.

Skin Regeneration Using Chimeric Implants

Epidermal substitutes were prepared for grafting when keratinocyte cultures had reached confluence as previously described.[22,23] They were grafted following the same procedure described in chapter 5. Chimeric epidermal substitutes were grafted onto Balb/c or C3H/HeN recipient mice depending on the syngeneic murine keratinocyte source (Fig. 6.3). As control experiments, homogeneous (100%) Balb/c and human epidermal sheets were also transplanted as isografts and xenografts, respectively. Five days later, macroscopic observations of the implants began. In all experiments, up to 11 days after grafting, cultured xenografts and chimeric (50%Balb/c-50%h, and 25%Balb/c-75%h) implants survived as well as isografted epidermal substitutes. No signs of necrotic zones were seen, as judged by visual inspection. Thirteen days post-grafting, xenogeneic implants presented necrotic zones, nevertheless chimeric implants showed the same macroscopic morphology as the isografts. Fourteen to 20 days after grafting, macroscopic observations showed minor and small necrotic zones. The percentage of skin replacement using chimeric epidermal substitutes was evaluated at 80 to 90% compared with 100%

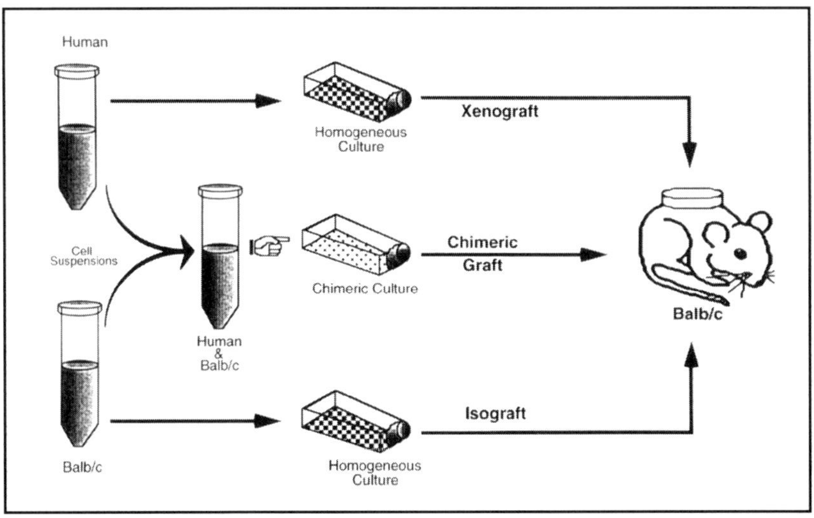

Fig. 6.3. Diagram showing homogeneous and chimeric (syngeneic-xenogeneic) epidermal substitute production and grafting

for isografts. Subsequent macroscopic observations showed total surface coverage and permanent skin replacement when the chimeric implants were used compared with the isografts (Fig. 6.4). These interesting observations need to be confirmed histologically.

STRUCTURAL ANALYSES OF THE NEWLY GENERATED SKIN

Histological studies were done of the newly generated cutaneous tissue after isografts, xenografts and chimeric (50%Balb/c-50%h and 25%Balb/c-75%h) epidermal substitute grafting. They were processed on days 14 and 30 using 4 µm-thick sections that were stained with hematoxylin and eosin. These analyses revealed a well-organized epidermis after chimeric grafting (Fig. 6.5). By contrast, histological

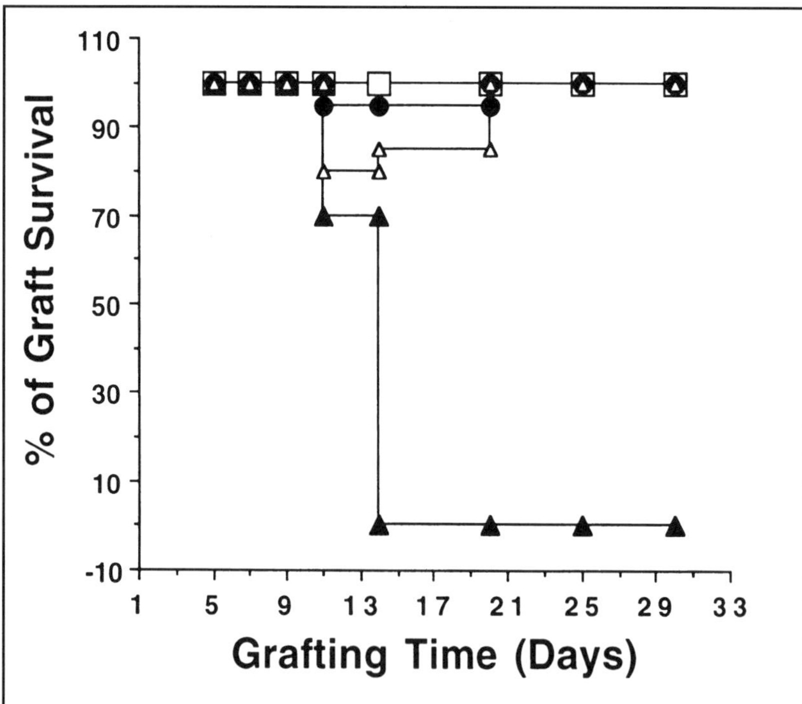

Fig. 6.4. Graft survival after chimeric grafting. Homogeneous Balb/c epidermal substitutes were produced and used for isografts (o), and human epidermal substitutes were used for xenografts (▲). Similarly, 50%B-50%h (●) and 25%B-75%h (Δ) chimeric graftable sheets were produced and grafted onto Balb/c adult recipients. Reprinted with permission from Rouabhia M, Transplantation 1996; 61(9):1290-1300.

studies of the xenografts showed complete degradation of the implant confirming the rejection symptoms that had been previously assessed by macroscopic observations. These important histological results were confirmed at 30 days after grafting (Fig. 6.5). The newly generated cutaneous tissue using chimeric sheet grafting was composed of basal and suprabasal cell layers and was comparable to tissue obtained with the isografts. Thus, the questions were:

> Do these newly generated cutaneous tissues secrete skin proteins such as laminin and type-IV collagen that are essential to the integrity of a healthy dermal-epidermal junction?
>
> Does the newly generated epidermis contain both xenogeneic and syngeneic keratinocytes?

Laminin and Type IV Collagen Syntheses

Thirty days after chimeric epidermal substitute grafting, biopsy specimens were harvested and labeled with anti-laminin and anti-type IV collagen specific antibodies. They showed positive staining

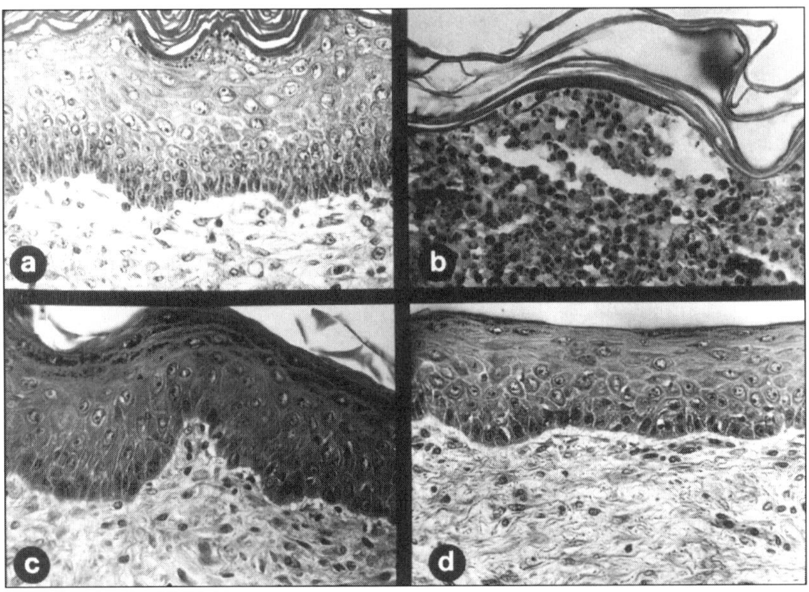

Fig. 6.5. Histological aspects of the newly generated epidermis 14 days after chimeric epidermal substitute grafting. Biopsies were performed 14 days postgrafting for histological studies: (a,e) isograft histology; (b,f) xenograft histology; (c,g) 50%B-50%h and (d,h) 25%B-75%h chimeric implant histologies. Magnification X300. Reprinted with permission from Rouabhia M, Transplantation 1996; 61(9):1290-1300.

for laminin (Fig. 6.6) and type IV collagen (Fig. 6.7). The presence of these proteins in the newly generated cutaneous tissue confirmed a well-structured skin with a dermal-epidermal junction. These histological and immunohistochemical structures obtained with chimeric implants were comparable to those seen with the isografts suggesting a well organized and functional cutaneous tissue.

Keratinocyte Phenotyping

Fourteen days after chimeric sheet grafting, the keratinocyte phenotypes were determined by an immunofluorescence staining method using specific antibodies (anti-HLA-ABC, anti-$H2^d/H2^k$). As shown in Figure 6.8, epidermis generated after 50%B-50%h chimeric epidermal substitute grafting contained only syngeneic keratinocytes such as the isografts 14 days after grafting. However, epidermis generated after 25%B-75%h chimeric implantation showed the presence of a reduced number of human keratinocytes. Furthermore, all

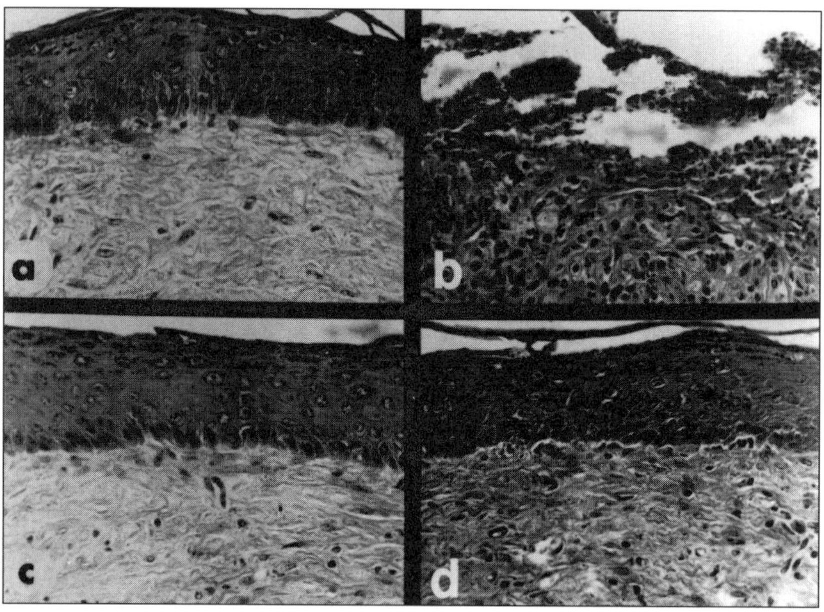

Fig. 6.6. Histological aspects of the newly generated epidermis 30 days after chimeric epidermal substitute grafting. Biopsies were performed one month later for histological studies: (a,e) isograft histology; (b,f) xenograft histology; (c,g) 50%B-50%h and (d,h) 25%B-75%h chimeric implant histologies. Magnification X300. Reprinted with permission from Rouabhia M, Transplantation 1996; 61(9):1290-1300.

remaining human cells in the newly generated epidermis were located in the upper layers of the epidermis suggesting that these cells were in the late stage of complete rejection. Thirty days after grafting the immunofluorescence staining method showed that xenogeneic keratinocytes were no longer present in the newly generated cutaneous tissue (data not shown). At both 14 and 30 days after chimeric grafting, xenogeneic keratinocytes seemed to be passively eliminated without acute rejection of full chimeric implants.

The previous analyses showed that complete xenogeneic grafts were rejected and xenogeneic keratinocytes present in the chimeric epidermal substitutes seemed to be selectively discarded from all chimeric implants. Thus the question arises: Are these xenogeneic keratinocytes eliminated via a cellular and humoral immune activation?

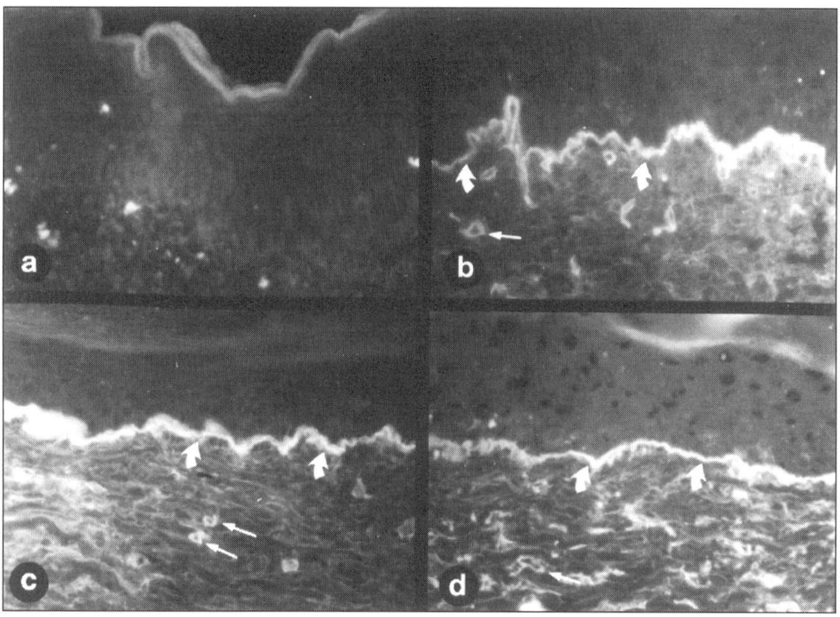

Fig. 6.7. Laminin synthesis by the newly generated cutaneous tissue. Immunofluorescence staining with anti-laminin monoclonal antibody of biopsy specimens taken 30 days after grafting showed a linear deposition of immunoreactants along the dermo-epidermal junction (curved arrows) and around dermal vasculature (straight arrows) after 50%B-50%h (c) and 25%B-75%h (d) chimeric implants. These laminin stainings in both chimeric implants were comparable to those obtained with (b) the isografts. The negative control is shown in panel (a). Magnification X300.

IMMUNOLOGICAL STUDIES

Cellular Response

Infiltration of the graft by mononuclear cells, producing a destructive lesion, is common in allograft and xenograft rejection.[25-27] The presence of white cell infiltrations was assessed using hematoxylin and eosin-stained chimeric biopsy specimens. These analyses showed (Fig. 6.9) an important leukocyte infiltration of the epidermal substitute xenografts, indicating the implant rejection. However, leukocyte infiltration was present in a small number in the chimeric (syngeneic-xenogeneic) implants 14 days after grafting, allowing well-structured skin replacement. Thirty days after grafting, the leukocytes infiltrating the chimeric implant were still present (Fig. 6.9), but seemed to be reduced in numbers. What is the phenotype of these infiltrating leukocytes? The answer to this question is shown in

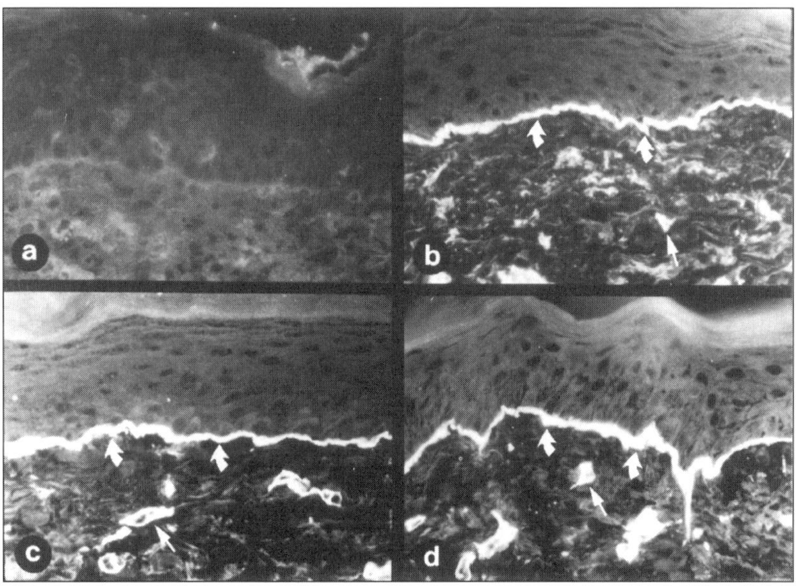

Fig. 6.8. Type IV collagen synthesis by the newly generated cutaneous tissue. Immunofluorescence staining with anti-type IV collagen polyclonal antibody of biopsy specimens taken 30 days after grafting, showed a linear deposition of immunoreactants along the dermal-epidermal junction (curved arrows) and around dermal vasculature (straight arrows) after 50%B-50%h (c) and 25%B-75%h (d) chimeric implants. These type IV collagen stainings in both chimeric implants were comparable to those obtained with (b) the isografts. The negative control is shown in panel (a). Magnification X300.

Figure 6.10. This infiltrate contained a large number of monocytes (Mac-1$^+$) in the xenografts, while a weak monocyte infiltration was noted in the chimeric implants. Using specific T cell antibodies, immunofluorescence staining revealed that the leukocyte infiltrate

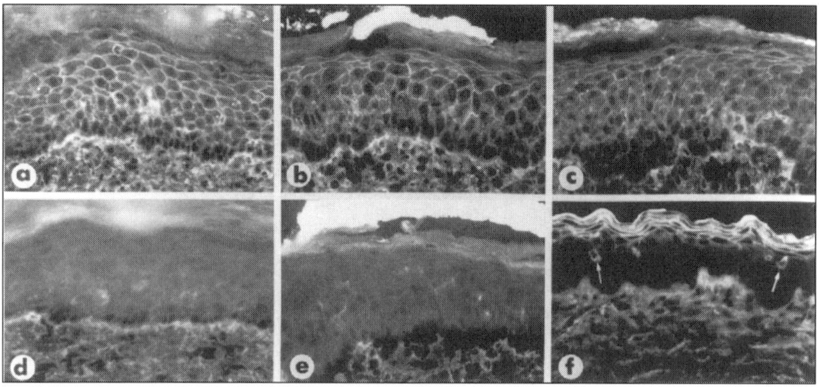

Fig. 6.9. Keratinocyte phenotyping in the newly generated cutaneous tissue after chimeric epidermal substitute grafting. Biopsy specimens were taken from the isografts and both chimeric implants 14 days after grafting, then stained with (a,b,c) anti-H2D^d or (d,e,f) anti-HLA-ABC-specific antibodies. (a,d) isografts; (b,e) 50%B-50%H chimeric implants; (c,f) 25%B-75%h chimeric implants. Only 25%B-75%h chimeric biopsy specimens (f) stained with anti-HLA-ABC revealed the presence of few human keratinocytes (arrows) located on the upper layers of the newly generated cutaneous tissue. Magnification X300. Reprinted with permission from Rouabhia M, Transplantation 1996; 61(9):1290-1300.

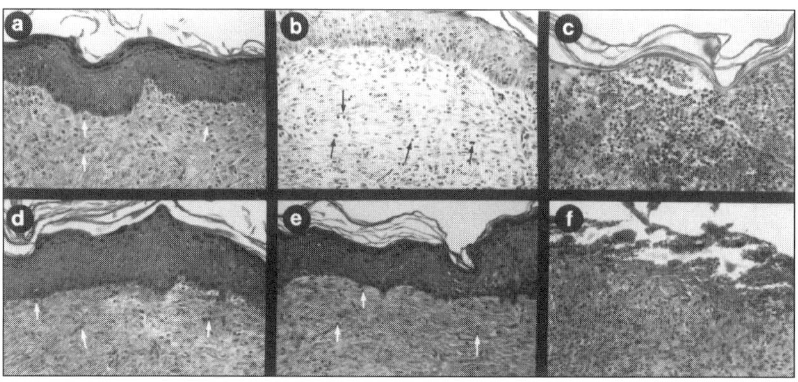

Fig. 6.10. Presence of infiltrating leukocytes in the newly generated cutaneous tissue after chimeric sheet grafting. Biopsies were performed 14 (a,b,c) and 30 (d,e,f) days after chimeric and xenogeneic sheet grafting for infiltrating leukocyte assessment using hematoxylin and eosin staining (arrows). Infiltrating cells in (a,d) 50%B-50%h, and (b,e) 25%B-75%h chimeric implants compared with (c,f) the xenografts (c,f). Magnification X150. Reprinted with permission from Rouabhia M, Transplantation 1996; 61(9):1290-1300.

contained CD4⁺ (Fig. 6.11) and CD8⁺ (Fig. 6.12) T cells in the xenograft as well as in the chimeric implants. The presence of infiltrating leukocytes was also found in the chimeric implant 30 days after grafting; their numbers seemed to be weaker than those obtained 14 days after grafting. The presence of leukocyte infiltration in the chimeric implant suggests cellular immunity activation. This cellular involvement could be followed by humoral immunity activation.

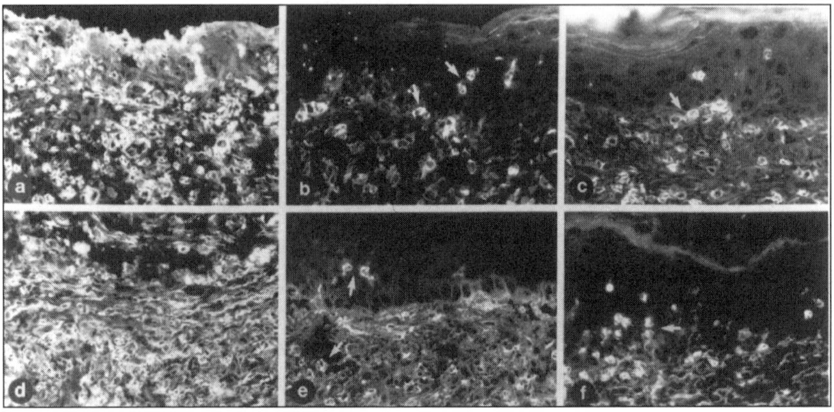

Fig. 6.11. Monocyte identification in the cellular infiltrate. Fourteen (a, b, c) and 30 (d, e, f) days after grafting, biopsy specimens were stained with anti-Mac-1 monoclonal antibody (arrows). Mac-1+ cells in (b, e) 50%B-50%h, and (c, f) in 25%B-75%h chimeric implants compared with (a,d) the xenografts. Magnification X300. Reprinted with permission from Rouabhia M, Transplantation 1996; 61(9):1290-1300.

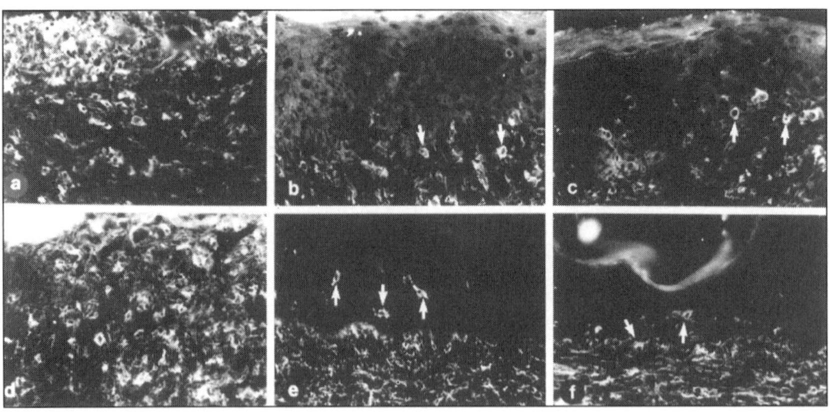

Fig. 6.12. Helper T cell identification in the cellular infiltrate. Fourteen (a,b,c) and 30 (d,e,f) days after grafting, biopsy specimens were stained with anti-CD4 monoclonal antibody (arrows). CD4+ cells in (b,e) 50%B-50%h, and (c,f) in 25%B-75%h chimeric implants compared with the (a,d) xenografts. Magnification X300. Reprinted with permission from Rouabhia M, Transplantation 1996; 61(9):1290-1300.

Humoral Response

Xenograft rejection is initiated through cellular immunity as well as through humoral immunity.[28-30] Consequently, signs of humoral immunity in xenogeneic cell rejection may occur after chimeric sheet grafting. To clarify this situation, sera were collected from isograft, xenograft and chimeric (50%Balb/c-50%h, 25%Balb/c-75%h) implant recipients 14 and 30 days after grafting. Frozen sections of normal human skin were stained with the previously harvested sera. Following this staining, sections were examined under a fluorescence microscope and photographed. Presented in Figure 6.13, the results reveal the presence of antibodies reacting with the human keratinocytes (epidermis). This positive staining showed the presence of circulating antibodies specifically reacting with xenogeneic keratinocytes.

Cytotoxic Activity of Chimeric Recipient Sera

Collected sera from isograft, xenograft and chimeric (50%Balb/c-50%h, 25%Balb/c-75%h) implant recipients 14 and 30 days after grafting were heat inactivated for 30 min at 56°C in a water bath and tested for the presence of anti-xenogeneic keratinocyte cytotoxic antibody using a complement-dependent cytotoxic assay. Human keratinocytes previously isolated and maintained in culture were suspended in a RPMI-1640 medium supplemented with 0.3% fetal calf serum.

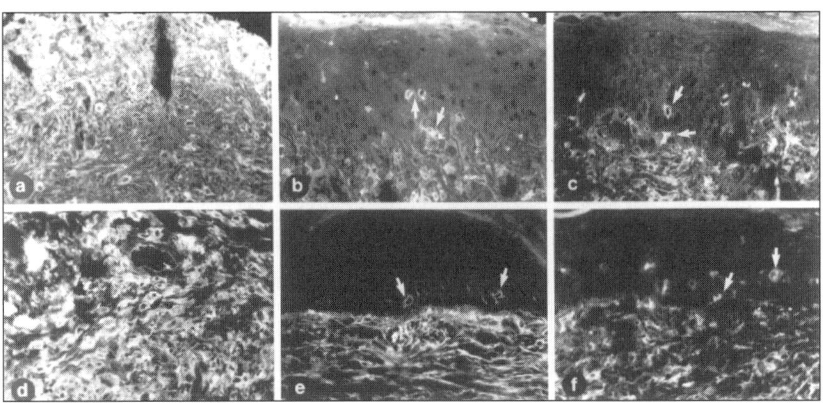

Fig. 6.13. Cytotoxic T cell identification in the cellular infiltrate. Fourteen (a,b,c) and 30 (d,e,f) days after grafting, biopsy specimens were stained with anti-CD8 monoclonal antibody (arrows). CD8+ cells in (b, e) 50%B-50%h, and (c, f) in 25%B-75%h chimeric implants compared to (a,d) the xenografts. Magnification X300. Reprinted with permission from Rouabhia M, Transplantation 1996; 61(9): 1290-1300.

Keratinocytes were incubated alone or with chimeric or homogeneous epidermal substitute recipient serum (25 μl) for 60 min at 4°C. The cells were then centrifuged and the supernatants discarded. Cells were incubated with Low-Tox rabbit complement (1/3) for 60 min at 37°C. Live and dead keratinocytes were scored using the trypan blue exclusion technique. More than 2000 cells were counted from each sample and the cytotoxic index (C.I.) was calculated.[30] According to Figure 6.14, chimeric sera showed cytotoxic activity on human keratinocytes. However, sera collected from xenografted mice showed higher cytotoxic activity than the chimeric sera. Serum harvested after 50%B-50%h chimeric sheet grafting seems to have a less cytotoxic effect than serum harvested after 25%B-75%h chimeric sheet grafting. Taken together, immunofluorescence staining and the cytotoxic index assessment showed humoral immunity activation. This response, coupled with cellular immunity, results in the selective elimination of the majority if not of all of xenogeneic keratinocytes present in the chimeric implants.

RELEVANCE OF THIS CHIMERIC CULTURE METHOD

Morphological, histological and immunohistochemical studies demonstrate that graftable chimeric (xenogeneic-syngeneic) epidermal substitutes allow permanent skin replacement. The newly generated cutaneous tissue was well structured and functional. These studies confirmed those results published previously,[30] and strengthen the feasibility of the chimeric cultured method on functional skin regeneration. As pointed out in chapter 5, clinical use of this chimeric culture method is an important improvement to burn treatment because:

1. it shortens the delay necessary to cover large cutaneous wounds;
2. before their elimination, allogeneic as well as xenogeneic cells present in the chimeric implant stimulate the autologous keratinocytes to proliferate and permanently cover the wounded bed;
3. this chimeric culture method reduces the cost of patient management during hospitalization and because
4. this chimeric culture method could be used for other therapies such as skin ulcers, congenital nevis, etc.

From a fundamental point of view, our studies showed that after chimeric epidermal substitute grafting, the newly generated skin essentially contained syngeneic keratinocytes suggesting the

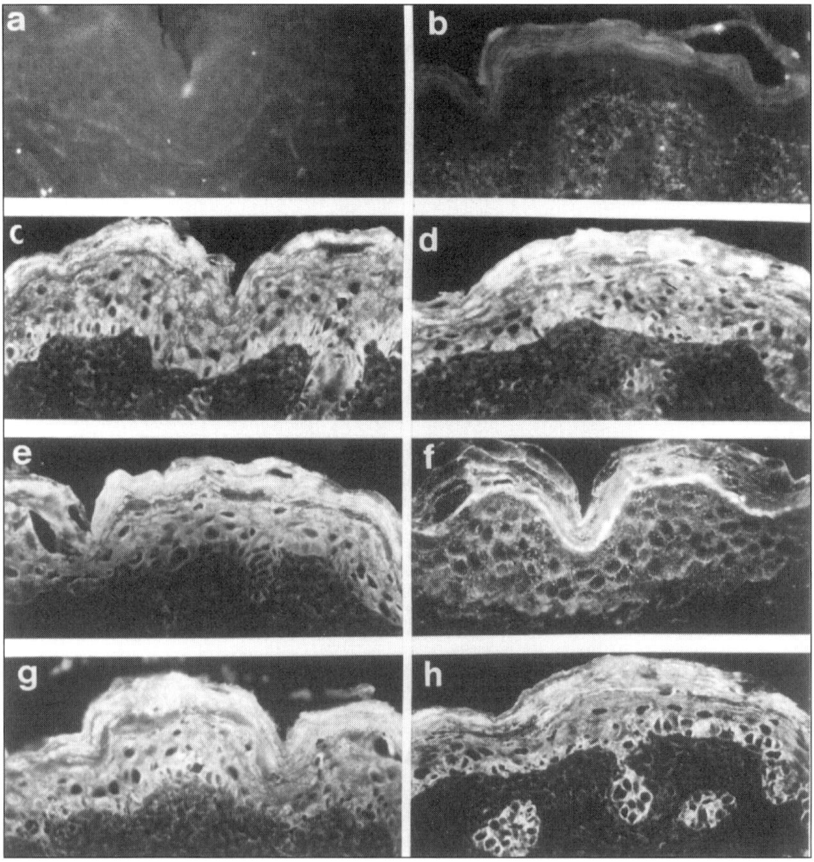

Fig. 6.14. Specific staining of human epidermis with chimeric recipient sera 14 and 30 days after grafting. Sera were collected from (e,f) 50%B-50%H and (g,h) 25%B-75%h chimeric sheet grafting and used to stain human skin biopsy specimens. Control sera were also collected from (a,b) isografts and (c,d) xenografts. Magnification X300. Reprinted with permission from Rouabhia M, Transplantation 1996; 61(9):1290-1300.

involvement of the recipient immune system. As indicated in the literature,[31-33] our immunological studies have revealed leukocyte infiltration of the xenografts and the chimeric implants. The presence of these white cells in the implants signified a specific cellular immunity activation. Although the recipient cellular and humoral immunity were activated, the rejection of xenogeneic cells initially present in the chimeric implant was passive. Indeed, the rejecting mechanisms initiated by the recipient were selective, resulting in a discrimination between "self" and "nonself" epidermal cells present in each chimeric

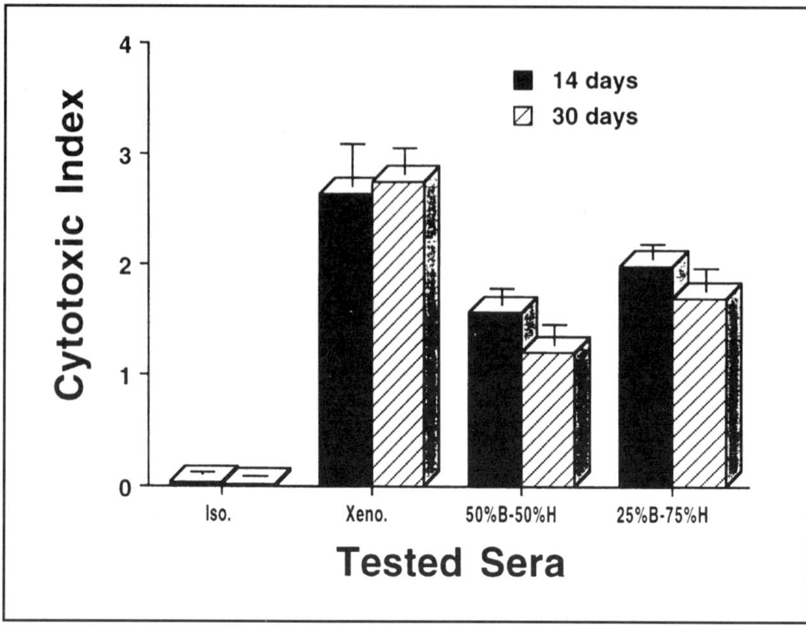

Fig. 6.15. Cytotoxic index of chimeric recipient sera at 14 and 30 days after grafting. Previously collected sera from isograft, xenograft, 50%B-50%h and 25%B-75%h chimeric implant recipients were used to evaluate their cytolytic capacity on xenogeneic keratinocyte cell suspensions. Results were mean percentage ± SD. Reprinted with permission from Rouabhia M, Transplantation 1996; 61(9):1290-1300.

implant, as demonstrated by the absence of xenogeneic keratinocytes in the newly generated cutaneous tissue. However, much needs to be learned about xenogeneic cell function (within the chimeric implant) before their selective elimination.

Meanwhile, it is well documented that during allograft and xenograft rejection,[34,35] T cell activation leads to a complex series of immune events. $CD4^+$ helper cells play a major role by finely regulating these two arms of the immune response through T and B cell cooperation and cytokine production. Because the complexity of the skin as an organ presents difficulties in the study of rejection and tolerance phenomena, the use of these chimeric cutaneous implants may help explain some of the immune responses involved in allogeneic and xenogeneic cutaneous tissue rejection or acceptance.

In summary, these chimeric (50%-50% and 25%-75%) cultured epidermal substitutes allow permanent functional skin replacement.[30,36] This represents a major improvement for massive damaged skin replacement because it significantly shortens the patient's waiting time for wound covering. This method could be much more effective if we are able to generate permanent skin using chimeric epidermal substitutes containing less than 25% autologous keratinocytes. Studies are presently in progress to evaluate this possibility.

Acknowledgments

The author thanks Julie Bergeron and Geneviève Ross for their excellent technical assistance and the members of the audiovisual service of Saint-Sacrement Hospital, Quebec. This work was supported by the MRC and the FRSQ.

References

1. Reemtsma K, McCracken BH, Schlegel JU et al. Heterotransplantation of the kidney: two clinical experiences. Science 1964; 143:700-702.
2. Auchincloss Jr H. Xenogeneic transplantation: a review. Transplantation 1988; 46:1-20.
3. Kaufman CL, Gaines BA, Ildstad ST. Xenotransplantation. Annu Rev Immunol 1995; 13:339-367.
4. Sandrin MS, Vaughan HA, Dabkowski et al. Anti-pig IgM antibodies in human serum react predominantly with Galα (1,3)Gal epitopes. Proc Natl Acad Sci USA 1993; 90:11391-11395.
5. Parker WR, Bruno D, Holzknecht ZE et al. Xenoreactive natural antibodies: isolation and initial characterization. J Immunol 1994; 153:3791-3803.
6. Batten P, Heaton T, Fuller-Espie S et al. Human anti-mouse xenorecognition. J Immunol 1995; 155:1057-1065.
7. Yamada K, Sachs DH, DerSimonian H. Human anti-porcine xenogeneic T cell response. Evidence for allelic specificity of mixed leukocyte reaction and for both direct and indirect pathways of recognition. J Immunol 1995; 155:5249-5256.
8. Davis ES, Pruitt SK, Green PS et al. Inhibition of complement, evoked antibody, and cellular response prevents rejection of pig-to-primate cardiac xenografts. Transplantation 1996; 62(7):1018-1023.
9. Wrenshall LE, Cerra FB, Singh RK et al. Heparan sulfate initiales signals in murine macrophages leading to divergent biological outcomes. J Immunol 1995; 154:871-880.
10. Bromberg BE, Song IC, Mohn MP. Use of pigskin as a temporary biological dressing. Plast Reconstr Surg 1965; 36:80-90.
11. Wood M, Hale MW. The use of pigskin in the treatment of thermal burns. Am J Surg 1972; 6:720-723.

12. Klein L, Mericka P, Preis J. Clinical experience with skin xenografts in burned patients. In: Masellis M, Gunn SWA, eds. The Management of Burns and Fire Disasters: Perspectives 2000. Dordrecht: Kluwer Academic Publishers, 1995:343-345.
13. Harris NS, Compton JB, Abston S et al. Comparison of fresh, frozen and lyophilized porcine skin as xenografts on burned patients. Burns 1976; 2:71-75.
14. Ding YL, Pu SS, Wu DZ et al. Clinical and histological observations on the application of intermingled auto- and porcine-skin heterografts in third degree burns. Burns 1982; 9:381-386.
15. Frank DH, Wachtel T, Frank HA et al. Comparison of Biobrane, porcine, and human allograft as biologic dressings for burn wounds. J Burn Care Rehab 1983; 4:186-189.
16. Hobby JAE, Levick DL. Clinical evaluation of porcine xenograft dressings. Burns 1977; 4:188-192.
17. Click RE, Weisberg D, White J et al. Rejection of rat skin xenografts by either murine CD4 or CD8 T Cells. Transplantation 1994; 58(9):1020-1026.
18. Swain SL, Dutton RW, Schwab R et al. Xenogeneic human anti mouse T cell responses are due to the activity of the same functional T cell subsets responsible for allospecific and major histocompatibility complex restricted responses. J Exp Med 1983; 157:720-729.
19. Bach FH, Dalmasso AP, Platt JL. Xenotransplantation: a current perspective. Transplant Rev 1992; 6:193-210.
20. Darby CR, Morris PJ, Wood KJ. Evidence that long-term cardiac allograft survival induced by anti-CD4 monoclonal antibody does not require depletion of $CD4^+$ T Cells. Transplantation 1992; 54:483-490.
21. Click RE, Adelmann A, Sherburne C et al. Genetic control of allograft prolongation. Transplant Proc 1986; 18(14):846-850.
22. Rouabhia M, Germain L, Bélanger F et al. Optimization of murine keratinocyte culture for the production of graftable epidermal sheets. J Dermatol 1992; 19(6):325-334.
23. Rouabhia M, Germain L, Bélanger F et al. Cultured epithelium allografts: Langerhans cell and Thy-1^+ dendritic epidermal cell depletion effects on allograft rejection. Transplantation 1993; 56(2):259-264.
24. Rouabhia M, Jobin N, Doucet Jr R et al. $CD36^+$-dendritic epidermal cells: a putative actor in the cutaneous immune system. Cell Transplant 1994; 3(6):529-536.
25. Platt JL. A perspective on xenograft rejection and accommodation. Immunol Rev 1994; 141:127-149.
26. Ahn M, Chang E, Barker CF et al. Cellular requirements for pancreatic islet xenograft rejection. Transplant Proc 1995; 27:3302-3303.
27. Sakakibara N, Click RE, Sakakibara K et al. Unconventional lymphocytes involved in rejection of xenogeneic heart grafts. Lab Invest 1990; 62(4):481-486.

28. Platt JL. Xenotransplantation: recent progress and current perspectives. Curr Opin Immunol 1996; 8(5):721-728.
29. Parker W, Saadi S, Lin SS et al. Transplantation of discordant xenografts: a challenge revisited. Immunology Today 1996; 17:373-378.
30. Rouabhia M. Permanent skin replacement using chimeric epithelial cultured sheets comprising xenogeneic and sygeneic keratincoytes. Transplantation 1996; 61(9):1290-1300.
31. Pierson III RN, Winn HJ, Russel PS et al. Xenogeneic skin graft rejection is especially dependent on $CD4^+$ T cells. J Exp Med 1989; 170:991-996.
32. Auchincloss Jr H, Moses R, Conti D et al. Xenograft rejection of class I espressing transgenic skin is CD4 dependent and CD8 independent. Transplant Proc 1990; 22:2335-2336.
33. Wren SM, Wang SC, Thui NL et al. Evidence for early Th2 cell predominance in xenoreactivity. Transplantation1993; 56:905-911.
34. Murray AG, Khodadoust MM, Pober JS et al. Porcine aortic endothelial cells activate human T cells: direct presentation of MHC antigens and costimulation by ligands for human CD2 and CD28. Immunity 1994; 1:57-63.
35. Rollins SA, Kennedy SP, Chodera AJ et al. Evidence that activation of human T cells by porcine endothelium involves direct recognition of porcine SLA and costimulation by porcine ligands for LFA-1 and CD2. Transplantation 1994; 57:1709-1716.
36. Rouabhia M, Germain L, Bergeron J et al. Allogeneic-syngeneic cultured epithelia. A successful therapeutic option for skin regeneration. Transplantation 1995; 59(9):1229-1235.

CHAPTER 7

In Vitro Engineered Human Skin Substitutes for UVB Harmful Effect Assessment

Nadine Chouinard, Roger Leblanc and Mahmoud Rouabhia

INTRODUCTION

Although solar radiation is essential for life, long-term ultraviolet (UV) exposure paradoxically constitutes a significant hazard to many living forms. Skin and eyes are organs which are directly exposed to solar radiation and are, as a consequence, particular targets of UVR-induced damage.[1,2] It is widely recognized that there is an important deleterious effect of the ultraviolet radiation (UVR) component of natural sunlight on cutaneous cells.[3,4] Epidemiological evidence collected over more than two decades has unequivocally identified UVB radiation overexposure as a risk factor for basal and squamous cell cancers.[5,6] Recently, UVR was shown to play an important pathogenic role in certain types of malignant melanoma.[7-10] Efforts to increase scientific knowledge about the photobiologic effects of UV light on the skin are important and relevant to both prevention and therapeutic treatment of human skin diseases. In this chapter, evidence for the deleterious effects of UVB on skin integrity and functions is presented. Also, the impact of this cutaneous tissue integrity disturbance on human life is pointed out. In addition, we summarized our studies on in vitro engineered skin substitute production and its feasibility as a useful tool for a better understanding of harmful UVB irradiation effects on skin integrity and function.

Skin Substitute Production by Tissue Engineering: Clinical and Fundamental Applications, edited by Mahmoud Rouabhia.

THE SUN: THE NATURAL SOURCE OF ULTRAVIOLET RADIATION

A sunny day generally puts people in a very good mood. Indeed in the summer, all of us look forward to outdoor activities with family members, with friends or alone. These outdoor activities are mainly dependent on warm and sunny weather conditions. Uninformed persons (the vast majority of the world's population) consider the sun innocuous. In fact, for them, the sun simply provides life-sustaining heat, light and energy. The repercussions engendered by injudicious overexposure to sun irradiation are often dismissed, due to their late pathologic signs.[8-10]

The sun is the major source of UV radiation.[11,12] The broad spectrum and intensity of UV from the sun are due to the high temperature

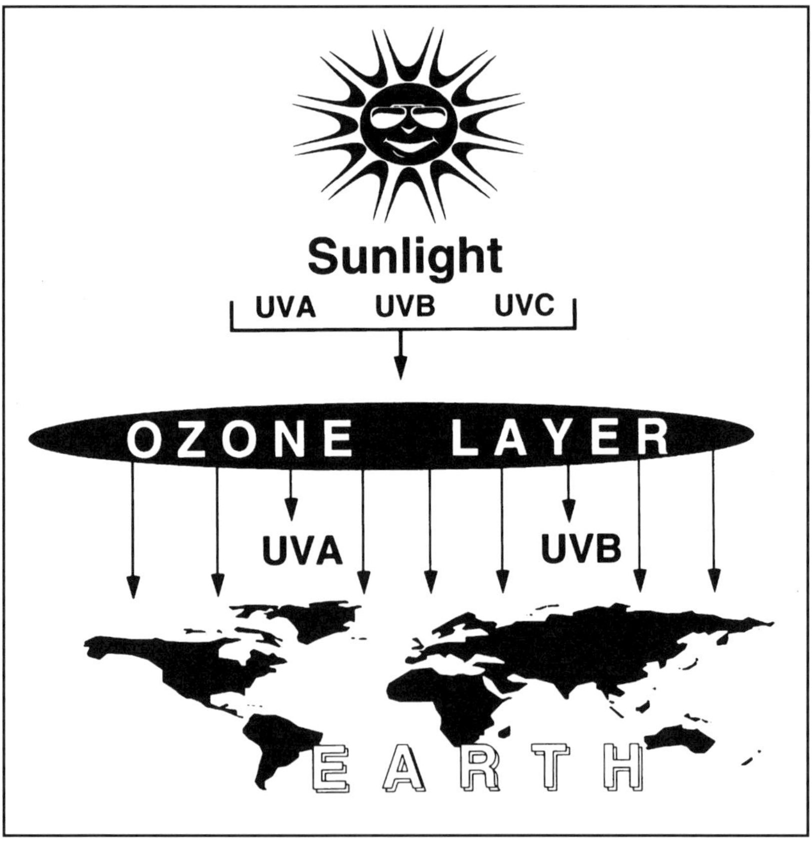

Fig. 7.1. Diagram summarizing UVR produced by the sun and reaching the earth.

on the earth's surface and its size. The intensity of solar UV reaching the earth is certainly highly dangerous and would probably be lethal to the majority if not all living organisms on the earth's surface without the shielding afforded by ozone. The spectrum of solar UV is divided into three categories (Fig. 7.1):

1. Ultraviolet-C (UVC) radiation ranges from 200 to 290 nm and is completely absorbed by the earth's ozone layers.[13] However, artificial sources (lamps for sterilization or disinfecting) may also emit UVC radiation.
2. Ultraviolet-B (UVB) radiation ranges from 290 to 320 nm.[13] A significant amount is absorbed by the ozone layer, and window glass attenuates it. This type of radiation causes most sunburn and redness to the skin.
3. Ultraviolet-A (UVA) radiation ranges from 320 to 400 nm.[13] As opposed to UVC and UVB, very little of this UVA radiation is absorbed by the ozone layer. The UVA dose reaching the earth is 100 to 1,000 times higher than UVB radiation.[14] In this chapter, we will only focus our interest on the effects of UVB radiation on the skin.

The amount of UVB that reaches various sites on the earth's surface varies significantly with latitude, altitude, season, time of day, cloudiness, reflectiveness of surfaces and atmospheric aerosols.[15] Thus, the amount of UVB an individual is exposed to is extremely dependent on the time, place and duration of exposure.

DELETERIOUS EFFECTS OF UVB ON THE SKIN

Ultraviolet radiation affects human skin continuously from birth to death. The concept that prolonged sun exposure is the most important environmental factor in the pathogenesis of skin cancer is not new. There are few, if any, healthy effects of exposure to UVB light. Nonprotected persons living in geographic regions of the world with the greatest amount (intensity, duration) of incidental sunlight have the highest rates of skin cancer. The highest distribution of malignant lesions is found on sun-exposed areas.[16] Indeed, the devastating effects of sunlight on the skin of farmers and sailors compared with that of indoor workers are presently well known. Morphologically, photophobes, avoiding the sun, have smooth, unblemished skin even in old age (80 to 90 years).[17] In contrast, photoaged skin displays visible disfigurements which reflect profound cutaneous structural changes.[18] Tissue shows a telangiectatic, leathery dry, nodular, yellow surface with a variety of benign, premalignant and malignant neoplasms, deep wrinkles and accentuated skin furrows.[19,20]

UVB incident on human skin can follow one of three courses: absorption, reflection, or scattering. Thus, UVB doses that reach the different skin layers (Fig. 7.2) are lower than the incident exposure.

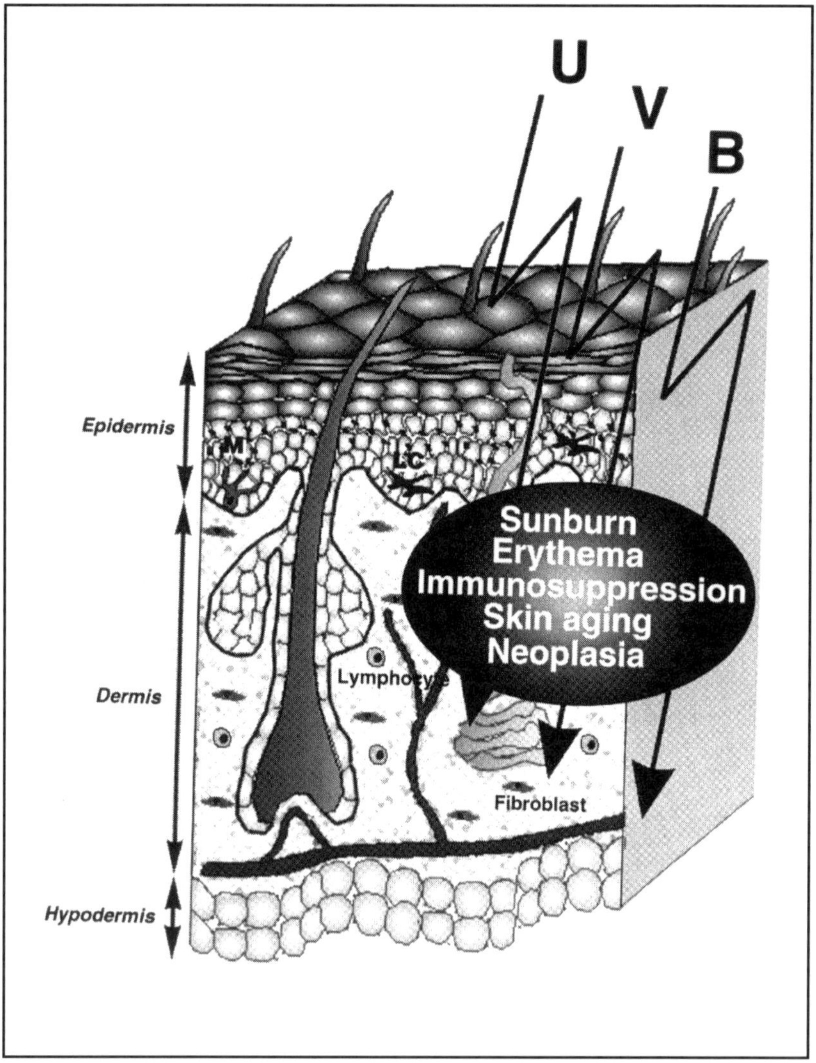

Fig. 7.2. Diagram showing UVB radiation reaching deep cutaneous layers.

Following this UVB penetration, a variety of cutaneous cells and structures are highly perturbed.[21-23]

UVB Consequences to the Skin

Clinically, skin exposure to UVB produces distinct patterns of inflammation. Acute exposure results in erythema, heat, edema, pain, and pruritus, followed by tanning and epidermal thickening.[24-26] Chronic exposure of human and animal to UVB radiation causes general and serious disturbances of all physiological systems of the body (Fig. 7.3) such as cataract formation, immunosuppression, etc. However, the significant pathological perturbations found in the skin are sunburn, actinic keratosis, skin aging and skin cancer.[27-31]

UVB Promotes Skin Cancer

The three main types of cutaneous malignancy are basal cell carcinoma, squamous cell carcinoma and malignant melanoma. The common incidence of these three types is highest malignancy. At the present time, the incidence of all types of skin cancer is rising. For all types of skin cancer, exposure to natural UV light (sun) is known to be a strong risk factor.[32,33] In the case of basal and squamous cell

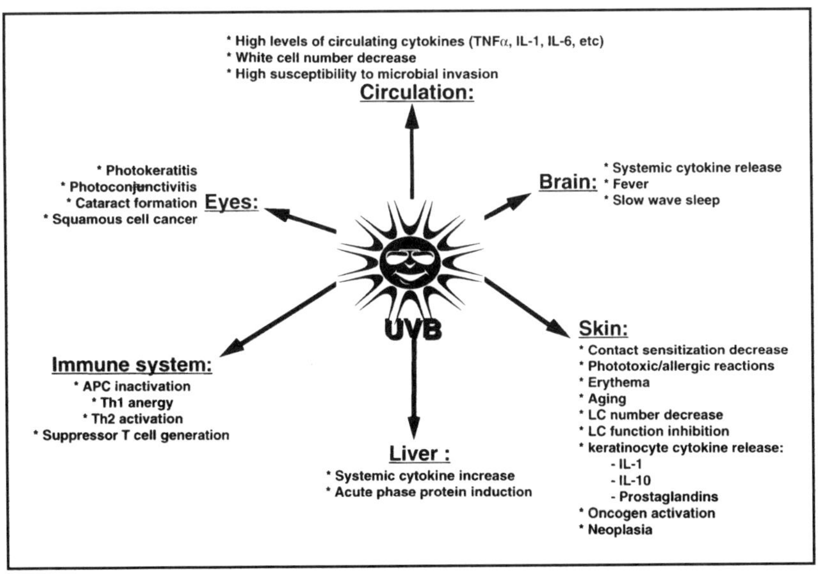

Fig. 7.3. Different perturbations induced by UVB on human organs and systems.

carcinoma, cumulative lifetime sun exposure appears to be the most important risk factor. Thus, patients with basal or squamous cell carcinoma are usually elderly and have the bulk of the lesions developing on chronically light-exposed skin such as the head and the neck area or the back of the hands. Indeed, morphohistological studies showed that cutaneous tissue alterations in sun-exposed areas of the face and hands are quite different from those in protected regions of aged persons. Thus, after five decades sun-exposed workers may look very old for their age with deep wrinkles and reduced skin elasticity. Following these morphohistological modifications, malignant cutaneous manifestations may occur.[34,35]

Data on UVB deleterious effects on the skin are generated using human volunteers for testing phototoxic agents such as UVB irradiation. Unfortunately, these evaluations are costly and difficult. To overcome these limitations, experimental models (rats, mice, etc.) are used for photodermatological studies.[36-38] Despite many similarities, the use of animal skin to elucidate mechanisms of UV-induced damage in human cutaneous tissues is hampered by inherent biological differences. Consequently, scientists are still looking for a cutaneous model that offers close similarity to human skin, ease of manipulation and is less costly.

To satisfy these three qualities, we engineered in vitro human skin substitutes[23] composed of healthy fibroblasts incorporated into collagen gel as an extracellular matrix (dermal substitutes = DS) and keratinocytes seeded onto these DS to obtain skin substitutes (SS). Using this model, we studied the effects of UVB radiation (10mJ/Cm^2) on extracellular matrix, on cutaneous cell morphology and functional activities.

DERMAL RESPONSES TO UVB STRESS

UVB Effects on In Vitro Produced Skin Substitutes

Following skin substitute production and irradiation (Fig. 7.4), biopsies collected from non-irradiated and irradiated SS showed that after irradiation the epidermal structure was composed of a large number of differentiating cells (faint nuclei and large cytoplasm). No healthy cuboidal cells even in the basal layer were seen as opposed to the epidermis of non-irradiated SS (Fig. 7.5f). Dermal cells were also

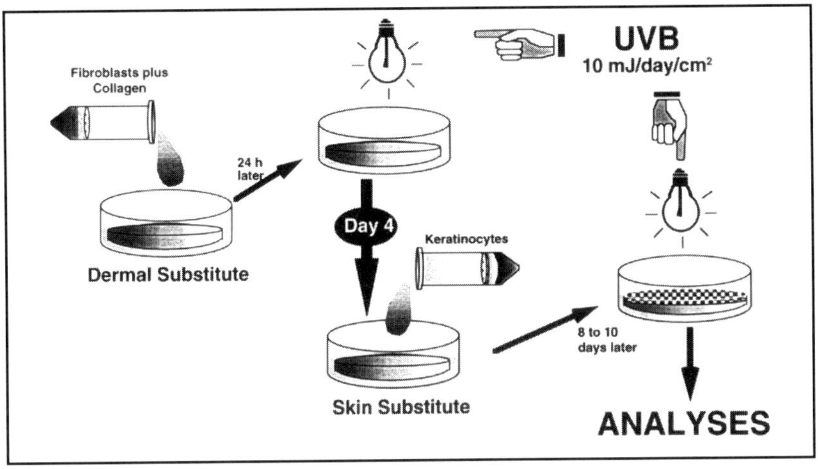

Fig. 7.4. Diagram summarizing different steps to in vitro produce skin substitutes and their usefulness in assessing UVB deleterious effects on skin components.

sensitive to UVB irradiation. Indeed, fibroblasts incorporated into collagen lattices as an extracellular matrix support were very bulky and presented a differentiated phenotype. The presence of keratinocytes on the DS did not prevent dermal fibroblasts from the deleterious effect of UVB radiation. Moreover, PAS staining showed (Fig. 7.6) that UVB radiation inhibited glycoprotein synthesis in the basement membrane zone. How does the extracellular matrix in our SS react to UVB irradiation?

Age-related changes in skin result from two distinct but concurrent processes: intrinsic and extrinsic aging.[39,40] The first is assumed to result from a genetically programmed sequence of events, whereas the second is dominated by dermal connective tissue disturbance that leads to the wrinkled, aged appearance of sun-exposed skin. These symptoms are due to cumulative injury induced by years of exposure to the sun. Photoaged dermis shows marked perturbations in the extracellular matrix components such as collagen and elastin.[41,42] Indeed, Fisher et al[43] showed recently that matrix-degrading metalloproteinase messenger RNAs, proteins and activities are induced in the skin within hours of exposure to UVB. The induction of proteins and proteolytic-enzyme syntheses occurred at UVB doses well below those that cause skin reddening. Within minutes, low-dose UVB upregulates the transcription factors AP-1 and NF-kB, which are known to be stimulators of metalloproteinase genes. These proteolytic enzymes may degrade extracellular matrix proteins (collagen and elastin) in the skin.

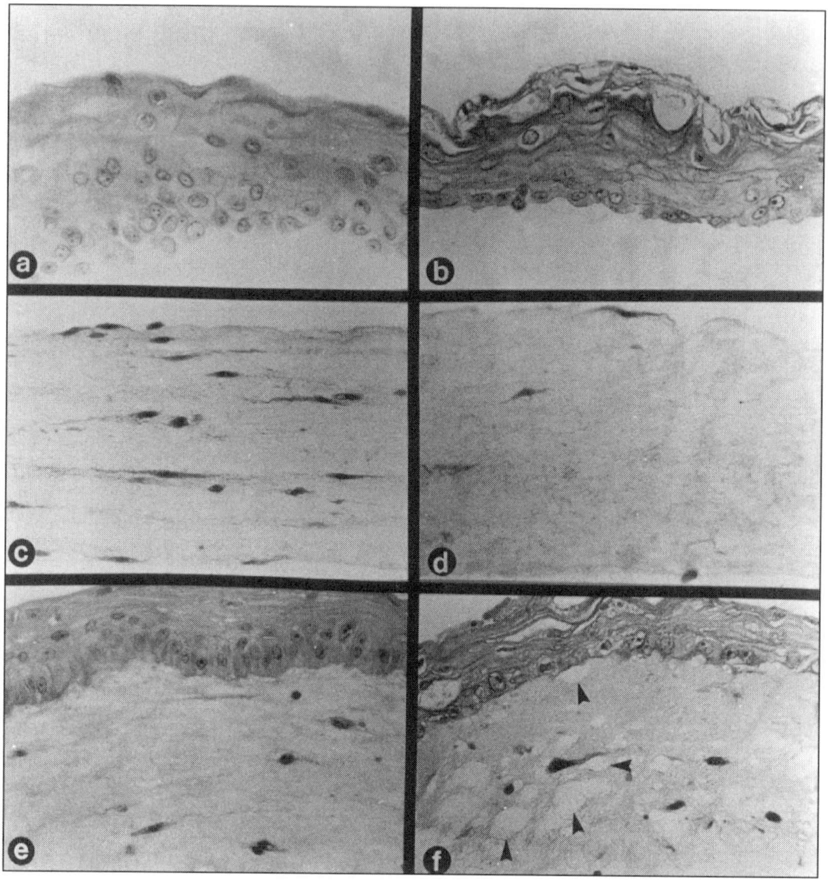

Fig. 7.5. Histological evaluations after UVB irradiation. Cells were cultured in monolayers or into and onto a collagen matrix, then irradiated (10mJ/cm2) daily. Non-irradiated (a,c,e) and irradiated (b,d,f) cells and a collagen matrix are represented. The deeper layer of the epidermal sheet was disorganized after UVB irradiation (b,f) as opposed to non-irradiated material (a,e). The collagen matrix structure was profoundly changed after UVB irradiation showing large holes (arrow heads). Finally, the presence of a reduced number of fibroblasts in the collagen matrix and differentiated keratinocytes in the monolayers or seeded onto the DS could be seen after UVB irradiation. Magnification X250. Reprinted with permission from Pâquet I et al, J Cell Physiol 1996; 166:296-304. Copyright ©1996 Wiley-Liss, Inc., a subsidiary of Wiley & Sons, Inc.

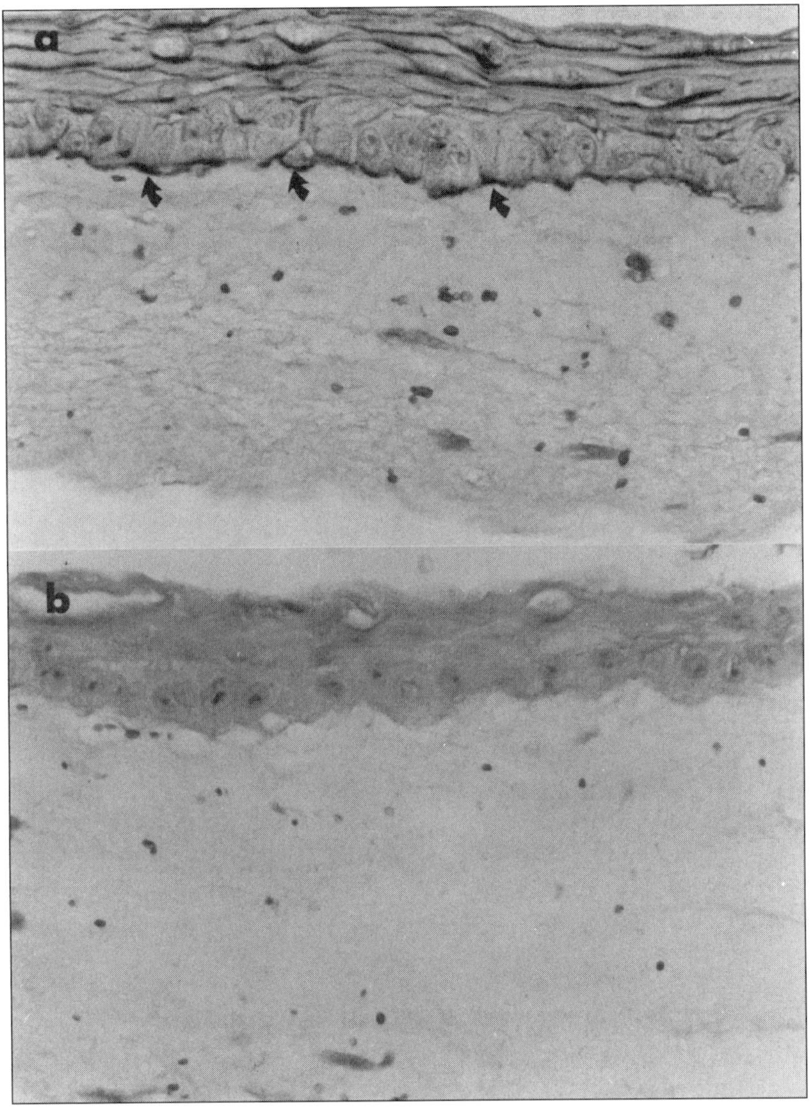

Fig. 7.6. Basement membrane zone alteration after UVB irradiation. Irradiated (10mJ/cm^2) and non-irradiated SS biopsies were stained with PAS and observed. (a) Non-irradiated SS biopsies and (b) irradiated SS biopsies. Note the specific staining of the dermal/epidermal zone on the non-irradiated SS compared to the UVB irradiated SS. Magnification X250. Reprinted with permission from Pâquet I et al, J Cell Physiol 1996; 166:296-304. Copyright ©1996 Wiley-Liss, Inc., a subsidiary of Wiley & Sons, Inc.

UVB Effects on Cutaneous Extracellular Matrix

Previous studies have shown that chronic exposure of the skin to UVB light leads to biochemical and ultrastructural changes in dermal collagen as the result of alterations in insoluble collagen and stable collagen cross-links.[44-48] Indeed, histopathologic analyses of sun-damaged skin have revealed reduced amounts of collagen. This reduction in collagen content may be due to decreased collagen synthesis or to increased collagen degradation. To these extracellular matrix morphological changes after UVB irradiation, we should add those obtained in our study using in vitro engineered skin substitutes. Histological studies using irradiated and non-irradiated unseeded collagen lattices showed that irradiated matrices presented large defects when compared to the non-irradiated controls. Indeed, histological studies revealed the presence of aggregated fibers and holes (Fig. 7.5f and Fig. 7.7). These results confirmed and strengthened those provided in the literature.

Elastin, which represents 2% of the dry weight of human skin, is also perturbed by UVB irradiation. This dermal protein is markedly increased (two to six times) in photoaged human and mouse skin. The massive accumulation of abnormal elastic fibers was noted in the upper- and mid-dermis, which appears to replace collagen fibers.[49-51] There is no change in the amino acid composition or immunologic reactivity of the elastin in sun-damaged skin that would explain why elastic fibers become thickened. However, the UVB mechanisms involved in elastin disturbance are still not known. Chronically sun-damaged human skin is characterized by the presence of an

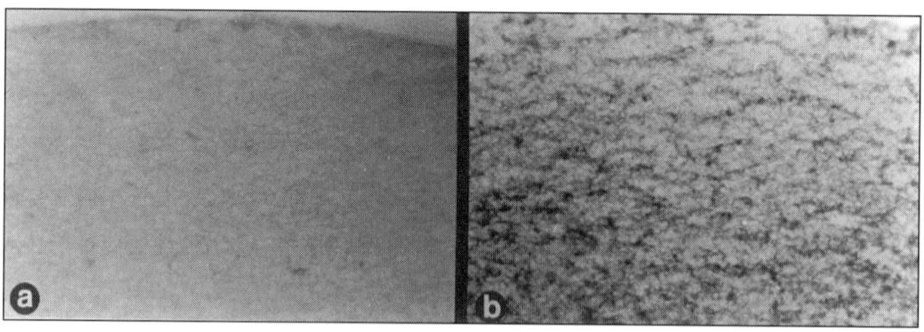

Fig. 7.7. Collagen matrix response to UVB irradiation. Collagen matrix free of cells was exposed or not to UVB irradiation (10mJ/cm²). (a) non-irradiated and (b) irradiated collagen matrix. Note the irregular collagen fiber arrangement (b) compared to the control (a). Magnification X250. Reprinted with permission from Pâquet I et al, J Cell Physiol 1996; 166:296-304. Copyright ©1996 Wiley-Liss, Inc., a subsidiary of Wiley & Sons, Inc.

inflammatory infiltrate.[5] This infiltrate contains various cytokines and proteases that may affect the dermal connective tissue.

UVB Effects on Resident Dermal Cells (fibroblasts)

Exposure of skin to UV irradiation is accompanied by alterations in dermal cellular (mast cells, endothelial cells, macrophages) metabolism resulting in the release of a series of mediators (histamine, arachidonic acid, PGE2, PGD2, 6-oxo-PGF1α and reactive oxidants). These mediators seem to profoundly modulate fibroblast functions.[52-55] In chronically irradiated skin, it has been shown that fibroblasts are larger than non-irradiated cells, with abundant cytoplasm.[56,57]

With engineered skin substitutes we evaluated some specific fibroblast parameters after UVB irradiation.[23] Morphological studies showed that fibroblast monolayer cultures were composed of large numbers of differentiating cells (Fig. 7.8d). These cells were characterized by their large size, wide and faint nuclei, large cytoplasm and the presence of vacuoles. No small proliferative cells were detected. This differentiating step was also observed in fibroblasts seeded into collagen (Fig. 7.8h). When non-irradiated fibroblasts reach confluence, they are morphologically small and cuboidal presenting dense nuclei and reduced cytoplasm. However, in irradiated cultures, cells never reached confluence due to the large number (if not all) of differentiating cells.

Quantitatively, these studies showed a significant decrease in the mitosis of fibroblasts seeded as monolayer cultures and in collagen matrices. This disturbance was mainly observed starting on day 4 of post-irradiation of fibroblast monolayers. Compared to non-irradiated cutaneous cells, irradiated fibroblasts generally reached only 70-80% confluence, while non-irradiated cells reached 100% confluence (Fig. 7.9b). The proliferative capacity of these fibroblasts was significantly decreased and may be completely inhibited by UVB irradiation. Decreased fibroblast density following UVB radiation raises a number of intriguing questions. One of these may relate to vimentin expression. For this purpose, biopsies were taken from irradiated and non-irradiated fibroblast monolayers, and SS, then stained with anti-vimentin monoclonal antibodies. These stainings showed (Fig. 7.10d,h) a higher fluorescence intensity in non-irradiated skin substitutes compared to irradiated ones.

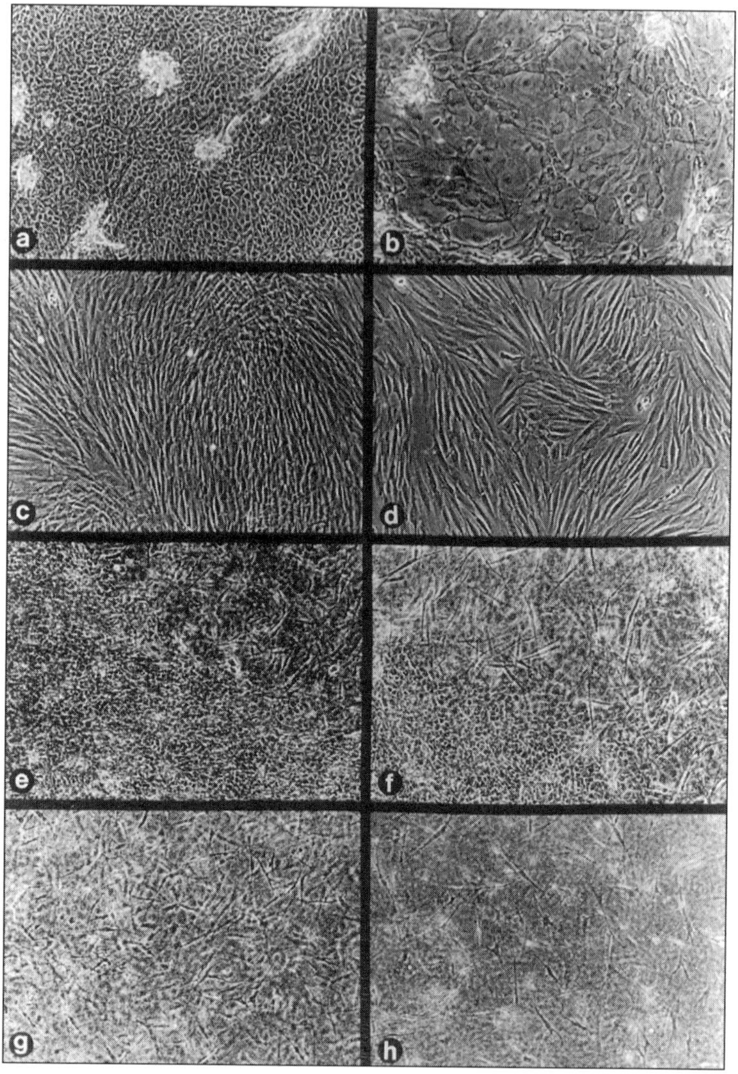

Fig. 7.8. Cutaneous cell morphology after UVB irradiation. Keratinocytes and fibroblasts were isolated from human normal skin biopsies and cultured in monolayers or incorporated into an extracellular matrix. (a,c) non-irradiated keratinocyte and fibroblast monolayers; (b,d) irradiated keratinocyte and fibroblast monolayers; (e,g) non-irradiated keratinocytes and fibroblasts seeded onto and into a collagen matrix respectively; (f, h) irradiated (10mJ/cm^2) keratinocytes and fibroblasts seeded onto and into a collagen matrix respectively. Note the cuboidal cells (a,e) and the regular morphology (a,c,e,g) of non-irradiated cells compared to differentiated (b,d) and bulky (b,d,f,h) cells after UVB irradiation. Magnification X250. Reprinted with permission from Pâquet I et al, J Cell Physiol 1996; 166:296-304. Copyright ©1996 Wiley-Liss, Inc., a subsidiary of Wiley & Sons, Inc.

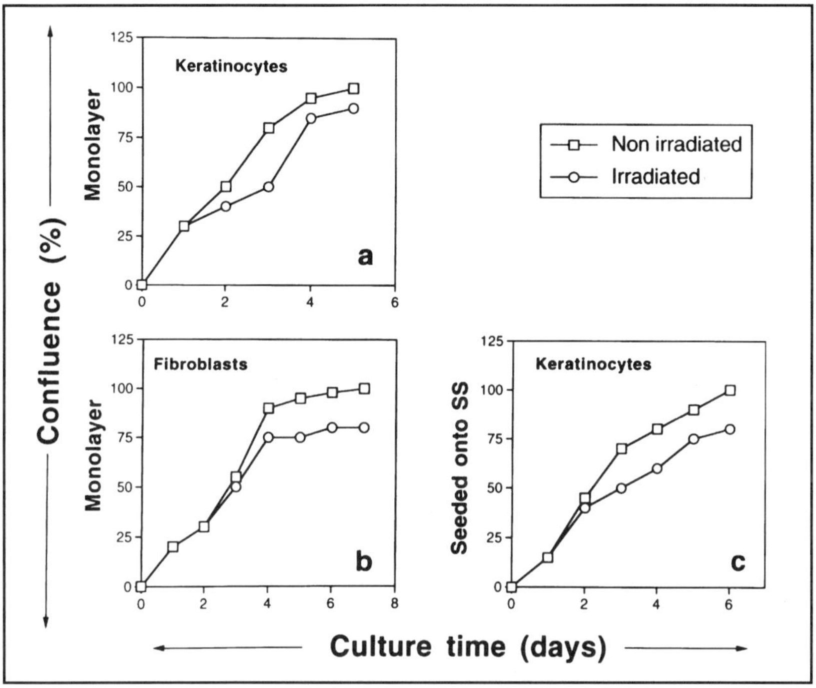

Fig. 7.9. Cell growth after UVB irradiation. Cutaneous cells were isolated and cultured. (a) Irradiated (10mJ/cm²) and non-irradiated keratinocyte monolayers, (b): Irradiated and non-irradiated fibroblast monolayers and (c): irradiated and non-irradiated keratinocytes seeded onto DS. Monolayer confluence was assessed then scored percentages were plotted. Non-irradiated cells (—□—); Irradiated cells (—○—). *, P< 0.05; **, P< 0.01 compared to non-irradiated materials. Reprinted with permission from Pâquet I et al, J Cell Physiol 1996; 166:296-304. Copyright ©1996 Wiley-Liss, Inc., a subsidiary of Wiley & Sons, Inc.

The present study and others[23,58,59] suggest that UVB induces morphological and functional disturbances in cutaneous cells. Also, it has been shown that UVB radiation induces functional disturbance in fibroblasts by decreasing their capacity to release growth factors such as EGF[56,60] that could explain the differentiating phenotype of UVB irradiated fibroblasts. However, little is known about molecular changes within the fibroblasts and the qualitative characteristics of their synthetic products after UVB stress. Indeed, it remains to be seen if elastic fibers, collagen and growing substances in irradiated skin are biochemically normal.

The microcirculation also suffers from sun exposure. Vessels become dilated and tortuous, seen on the surface as telangiectasias;

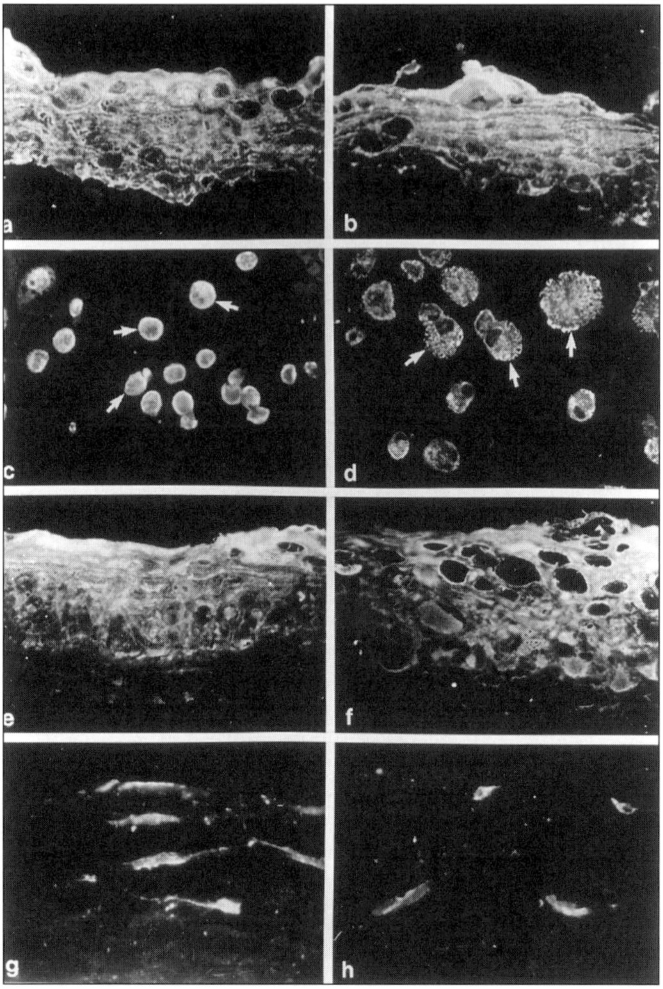

Fig. 7.10. Cytokeratin and vimentin staining after UVB irradiation. Irradiated (10mJ/cm^2) and non-irradiated biopsies were stained for cytokeratin-AE3 (keratinocytes) and vimentin (fibroblasts) expression. (a,b): keratinocyte monolayer, (c,d): fibroblast monolayer, (e,f): keratinocytes grown on the skin substitute, and (g,h): fibroblasts incorporated into collagen matrices. Non-irradiated (a,c,e,g) and irradiated (b,d,f,h) monolayer cells, and DS and SS are presented. Cytokeratin expression was decreased after UVB irradiation of monolayer keratinocytes (b) and those seeded onto the DS (f) compared to non-irradiated cells (a,e). Note the cell morphology and reduced vimentin expression of irradiated fibroblasts in monolayers (d) and those incorporated into the collagen matrix (h), compared to non-irradiated materials (c,g). Magnification X250. Reprinted with permission from Pâquet I et al, J Cell Physiol 1996; 166:296-304. Copyright ©1996 Wiley-Liss, Inc., a subsidiary of Wiley & Sons, Inc.

they also become very spare, with the horizontal superficial plexus almost completely destroyed.[61,62] Our SS model will certainly help to evaluate vessel endothelial cell modification after UVB irradiation and their control.

EPIDERMAL RESPONSES TO UVB

There are a number of light microscopic, ultrastructural and histochemical investigations concerning the injurious effects of UV radiation on keratinocytes of humans and animals.[63-65] One of the characteristic findings after UV radiation is the occurrence of individual cell keratinization "sunburn cells" which are found diffused throughout the epidermis.[66,67] Thickening of the stratum corneum is also a prominent change which appears after chronic UV exposure. These obvious alterations in cellular differentiation strongly suggest that UV affects epidermal keratinization. It has been shown that, following UVB irradiation, rapid alterations in the distribution pattern of keratins were observed in the epidermis. This change is a reflection of a hyperproliferative state of epidermal keratinocytes induced by accelerated DNA synthesis after UVB irradiation.

UVB Effects on Keratinocytes

Human keratinocytes are the most appropriate target cells for evaluating the cutaneous toxicity and pharmacology of chemicals or UV radiation. To this end, in vitro produced skin substitutes could be useful in the field of UV-induced skin damage.[58,68] Using this technology, monolayer and SS were produced, UVB irradiated and then analyzed. Morphologically, keratinocyte monolayers were composed of large numbers of differentiating cells (Fig. 7.8b). These cells were characterized by their large size, wide nuclei, large cytoplasm and the presence of vacuoles. No proliferative small cuboidal cells were detected. This differentiating step was also observed with keratinocytes of the SS (Fig. 7.5b). When non-irradiated keratinocytes reached confluence, they presented a small cuboidal morphology with a dense nucleus and reduced cytoplasm. However, in irradiated cultures, cells never reached confluence due to the large number (if not all) of differentiating cells. Functionally, UVB irradiation significantly disturbed keratinocyte proliferation (Fig. 7.9a,c). This growth disturbance was mainly observed starting on day 2 of post-irradiation in keratinocyte monolayers, and on day 4 in the keratinocyte layers of SS. Compared to non-irradiated cutaneous cells, irradiated keratinocytes were generally at less then 70% of confluence, while non-irradiated cells reached 100% confluence.

Histologically, irradiated epidermal substitutes were composed of a large number of differentiating cells (faint nuclei and large cytoplasm) which are present in each layer. No healthy cuboidal cells in the deep (basal) layer were seen as opposed to the non-irradiated epidermal substitutes. These histological alterations were confirmed with keratinocytes seeded onto DS (Fig. 7.8f). Moreover, UVB radiation inhibited glycoprotein synthesis in the basement membrane zone suggesting that UVB acts on the keratinocyte and fibroblast cooperation (Fig. 7.6). However, there was no significant effect on cytokeratin (AE_3) expression between irradiated and non-irradiated keratinocytes (Fig. 7.10a.b).

UVB Effects on Melanocytes

Epidermal melanocytes are the main contributor to skin pigmentation. They are continuously interacting with keratinocytes.[69,70] Indeed, melanocytes produce melanin pigment in melanosomes, which are then transferred and distributed into adjacent keratinocytes through extended dendritic processes in order to protect the skin against solar UVB. In response to UV light, there is an immediate redistribution of pre-existing melanin, followed by enhanced melanogenesis and increased transfer of melanin to epidermal keratinocytes. Within the last decade, homogeneous melanocyte culture systems became available to study melanogenesis in response to UV radiation in vitro.[71,72] These studies permitted demonstration of direct UV effects on melanocytes but precluded an investigation of paracrine effects due to epidermal keratinocytes and dermal fibroblasts. More recently, skin equivalent models have become available. They may offer convenient means to study in vitro cutaneous cell-cell interactions after UVB irradiation. Early attempts using dermal substitutes showed that UV irradiation stimulated melanogenesis and enhanced pigment transfer from melanocytes to keratinocytes.[73]

DELETERIOUS EFFECTS OF UVB ON CUTANEOUS IMMUNE SYSTEM

Cutaneous exposure to UVB radiation is associated with altered immune responses and impairs the induction of contact hypersensitivity to hapten applied directly to the irradiated skin. To explain these immunosuppression effects, several mechanisms have been intricated. Animal studies showed that UVB irradiation results in a reduction in Langerhans cell (LC) number and the functional alteration of those LC still in the skin. The state of this immunosuppression seems to be mediated by the generation of antigen-specific suppressor T cells.[74-77]

In humans, UVB induces the appearance in the epidermis of DR+/T6- cells, which are distinct from LC and are involved in the activation of T-suppressor circuit. LC in UVB irradiated skin are converts from an immunogenic to a tolerogenic antigen-presenting cell (APC). Consequently, there is imbalance between Th1 (mediating delayed-type hypersensitivity) and Th2 (tolerance) subsets. Furthermore, the UVB-induced immunosuppression goes through the synthesis and secretion of several soluble mediators and cytokines such as IL-1, TNF-α, etc.[78-79]

Although many investigations have been undertaken, there are still several questions to be answered, not only on the cellular and molecular levels but also regarding the mechanisms underlying the integrating activity of the immune system in general and, specifically, the cutaneous immune system after UVB irradiation. Once again, using in vitro engineered human skin substitute containing active leukocytes with fibroblasts and keratinocytes, we may be able to answer some relevant questions regarding the cutaneous immune system perturbation after UV irradiation.

CONCLUSION

Studies of the effects of UVB on the skin have generated a plethora of information in the fields of clinical and experimental dermatology, immunology, photochemistry and photobiology. As the skin is the main organ directly exposed to UVR, several questions regarding the structural and functional perturbation induced in the skin by UV irradiation were asked. Some of these questions were answered using human and animal studies. These answers were confirmed using engineered SS. Indeed, data obtained from keratinocyte and fibroblast monolayers, and DS and SS repeatedly exposed to UVB irradiation strongly support the hypothesis that such radiation significantly weakens the in vivo structural organization of dermal and epidermal cells as well as extracellular matrix components. Thus, normal skin integrity is at risk following UVB exposure, leading to a major loss of functional properties, notably as a barrier against infections and as a specific immune organ.

The use of skin substitutes sharing the same histological and functional features as real skin will allow a better understanding of the physiological events which lead to UV-related problems in humans, and is a promising tool for pharmacological and toxicological studies.

Acknowledgments

We are grateful to Julie Bergeron and Geneviève Ross for their excellent technical assistance, as well as to the members of the audiovisual service of The Saint-Sacrement Hospital. This work was supported by the Fondation de l'Hôpital du St-Sacrement and Hydro-Québec.

References

1. Kripke ML. Immunological mechanisms in UV irradiation carcinogenesis. Adv Cancer Res 1981; 34:69-106.
2. Ziegler A, Jonason S.A, Leffell JD et al. Sunburn and p53 in the onset of skin cancer. Science 1994; 372:773-776.
3. Longstreth J. Cutaneous malignant melanoma and ultraviolet radiation: a review. Cancer Metab Rev 1988; 7:321-333.
4. Scollo J, Fears TR. The association of solar ultraviolet and skin melanoma incidence among caucasians in the united states. Cancer Invest 1987; 5:275-283.
5. Streilein JW, Taylor JR, Vincek V et al. Relationship between ultraviolet radiation-induced immunosuppression and carcinogenesis. J Invest Dermatol 1994; 103:107s-111s.
6. Grabbe S, Granstein RD. Photoimmunology of UV-induced skin cancer. In: Krutman J, Elmets CA, eds. Photoimmunology. Oxford, UK: Blackwell Sciences Ltd, 1995:99-118.
7. Moan J, Dahlback A, Larsen S et al. Ozone depletion and its consequences for the fluence of carcinogenic sunlight. Cancer Res 1989; 49:4247-4250.
8. Vitasa BC, Taylor HR, Strickland PT et al. Association of nonmelanoma skin cancer and actinic keratosis with cumulative solar ultraviolet exposure in Maryland watermen. Cancer 1990; 65:2811-2817.
9. Oikarinen A, Peltonen J, Kallioinen M. Ultraviolet radiation in skin ageing and carcinogenesis: the role of retinoids for treatment and prevention. Ann Med 1991; 23:497-505.
10. Madronich S, de Gruijl FR. Skin cancer and UV radiation. Nature 1993; 366:23.

11. Taylor CR, Stern RS, Lyden JL et al. Photoaging/photodamage and photoprotection. J Am Acad Dermatol 1990; 22(1):1-15.
12. Young AR. Cumulative effects of ultraviolet radiation on the skin: cancer and photoaging. Semin Dermatol 1990; 9(1):25-31.
13. Odom RB. Focus on Photodamage: A Medical Problem. Parsippany, NJ: Skin Phototrauma Foundation 1990:1-29.
14. Frederick JE, Snell HE, Haywood EK. Solar ultraviolet radiation at the earth's surface. Photochem Photobiol 1989; 51:443-450.
15. Scotto J, Kopf AW, Urbach F. Nonmelanoma skin cancer in four areas of the U.S. Cancer 1974; 34:1331-1338.
16. Kligman AM. Early destructive effects of sunlight on human skin. JAMA 1969; 210:2377-2380.
17. Kligman LH. Skin changes in photoaging: characteristics, prevention, and repair. In: Balin AK, Kligman AM, eds. Aging and the Skin. New York: Raven Press, 1989:331-346.
18. Gilchrest BA, Soter NA, Hawk JLM et al. Histologic changes associated with ultraviolet A induced erythema in normal human skin. J Am Acad Dermatol 1983; 9:213-219.
19. Seguin LR, Ganges MB, Tarone RE et al. Skin cancer and chromosomal aberrations induced by ultraviolet radiation. Evidence for lack of correlation in xeroderma pigmentosum variant and group E patient. Cancer Genet Cytogenet 1992; 60:111-116.
20. Moan J, Dahlback A, Henriksen T et al. Biological amplification factor for sunlight-induced nonmelanoma skin cancer at high latitudes. Cancer Res 1989; 49:5207-5212.
21. Bruls WAG, Slaper H, Van der Leun JC et al. Transmission of human epidermis and stratum corneum as a function of thickness in the ultraviolet and visible wavelengths. Photochem Photobiol 1984; 40(4):485-494.
22. Wefers H, Melnik BC, Flür M et al. Influence of UV irradiation on the composition of human stratum corneum lipids. J Invest Dermatol 1991; 96:959-962.
23. Pâquet I, Chouinard N, Rouabhia M. Cutaneous cell and extracellular matrix responses to ultraviolet-B irradiation. J Cell Physiol 1996; 166:296-304.
24. Grewe M, Trefzer U, Ballhorn A et al. Analysis of the mechanism of ultraviolet (UV) B radiation-induced prostaglandin E_2 synthesis by human epidermoid carcinoma cells. J Invest Dermatol 1993; 101:528-531.
25. Parrish JA, Jaenicke KF, Anderson RR. Erythema and melanogenesis action spectra of normal human skin. Photochem Photobiol 1982; 36:187-191.
26. Hruza LL, Pentland AP. Mechanisms of UV-induced inflammation. J Invest Dermatol 1993; 100:35s-41s.
27. MacKie RM. Sun and skin damage. J Internal Medicine 1993; 234:111-113.

28. Garmyn M, Degreef H, Gilchrest BA. The effect of acute and chronic photodamage on gene expression in human keratinocytes. Dermatology 1995; 190:305-308.
29. Bowden GT, Nelson MA, Levy JP et al. Molecular mechanisms of skin carcinogenesis induced by chemicals and ionizing radiation. In: Hecker E, Jung EG, Marks F, Tilgen W, eds. Recent Results in Cancer Research. Berlin, Heidelberg: Springer-Verlag 1993:215-227.
30. Jeevan A, Kripke ML. Ozone depletion and the immune system. The Lancet 1993; 342:1159-1161.
31. Donawho CK, Kripke ML. Evidence that the local effect of ultraviolet radiation on the growth of murine melanomas is immunologically mediated. Cancer Res 1991; 51:4176-4181.
32. Armstrong BK. Implications of increased solar UVB for cancer incidence. In: Chamin ML, ed. The Role of the Stratosphere in Global Change. Berlin, Heidelberg, New-York: Springer-Verlag, 1993:517-540.
33. Holman CDJ, Armstrong BK. Pigmentary traits, ethnic origin, benign nevi, and family history as risk factors for cutaneous malignant melanoma. J Natl Cancer Inst 1984; 72:257-266.
34. Hughes BR, Cunliffe WJ, Bailey CC. Excess benign melanocytic naevi after chemotherapy for malignancy in childhood. Br Med J 1989; 299:88-91.
35. Gilchrest BA, Zhai S, Eller MS et al. Treatment of human melanocytes and S91 melanoma cells with the DNA repair enzyme T4 endonuclease V enhances melanogenesis after ultraviolet irradiation. J Invest Dermatol 1993; 101:666-672.
36. Cole CA. Animal models for sunscreens and photoprotection. In: Lowe NJ, Shaath NA, eds. Sunscreens: Development, Evaluation, and Regulatory Aspects. New York: Marcel Dekker, Inc., 1990:447-457.
37. Sambuco CP. Miniature swine as an animal model in photodermatology. Photodermatology 1985; 2:144-150.
38. Wolf P, Yarosh DB, Kripke ML. Effects of sunscreens and a DNA excision repair enzyme on ultraviolet radiation-induced inflammation, immune suppression, cyclobutane pyrimidine dimer formation in mice. J Invest Dermatol 1993; 101:523-527.
39. Braverman IM, Fonferko E. Studies in cutaneous aging. I. The elastic fiber network. J Invest Dermatol 1982; 78:434-443.
40. Gilchrest BA, Stoff JS, Soter NA. Chronologic aging alters the response to UV-induced inflammation in human skin. J Invest Dermatol 1982; 79:11-16.
41. Schwartz E. Connective tissue alterations in the skin of ultraviolet irradiated mice. J Invest Dermatol 1988; 91:158-161.
42. Montagna W, Carlisle K. Structural changes in aging human skin. J Invest Dermatol 1979; 73:47-53.
43. Fischer GJ, Datta SC, Talwar HS et al. Molecular basis of sun-induced premature skin ageing and retinoid antagonism. Nature 1996; 379:335-339.

44. Eyre DR, Paz MA, Gallop PM. Cross-linking in collagen and elastin. Annu Rev Biochem 1984; 53:717-748.
45. Smith JG, Davidson EA, Sams WM et al. Alteration of dermal connective tissue with age and chronic sun damage. J Invest Dermatol 1962; 39:347-350.
46. Uitto J. Connective tissue biochemistry of the aging dermis. Age-related alterations in collagen and elastin. Dermatol Clin 1986; 4:433-446.
47. Wlaschek M, Bolsen K, Herrmann G et al. UVA-induced autocrine stimulation of fibroblast-derived-collagenase by IL-6: a possible mechanism in dermal photodamage. J Invest Dermatol 1993; 101:164-168.
48. Yamauchi M, Prisayanh P. Haque Z et al. Collagen cross-linking in sun-exposed and unexposed sites of aged human skin. J Invest Dermatol 1991; 97:938-941.
49. Smith JG, Davidson EA, Clark RD. Dermal elastin in actinic elastosis and pseudoxanthoma elasticum. Nature 1962; 195:716-717.
50. Smith JG, Davidson EA, Sams Jr WM et al. Alterations in human dermal connective tissue with age and chronic sun damage. J Invest Dermatol 1962; 347-350.
51. Wrenn DS, Parks WC, Whitehouse LA et al. Identification of multiple tropoelastins secreted by bovine cells. J Biol Chem 1987; 262:2244-2249.
52. Diffey BL, Oakley AM. The onset of ultraviolet erythema. Br J Dermatol 1987; 116:183-187.
53. Hawk HLM, Black AK, Jaenicke KF et al. Increased concentrations of arachidonic acid, prostaglandins E_2, D_2 and 6-oxo-F_{14} and histamine in skin following UVB irradiation. J Invest Dermatol 1983; 80:494-499.
54. Gilchrest BA, Soter NA, Stoff JS et al. The human sunburn reaction: histologic and biochemical studies. J Am Acad Dermatol 1981; 5:411-422.
55. Pentland AP, Mahoney M, Jacobs SC et al. Enhanced prostaglandin synthesis after ultraviolet injury is mediated by endogenous histamine stimulation. J Clin Invest 1990; 86:566-574.
56. Petersen MJ, Hansen C, Craig S. Ultraviolet A irradiation stimulates collagenase production in cultured human fibroblasts. J Invest Dermatol 1992; 99:440-444.
57. Gaboriau F, Morlière P, Marquis I et al. Membrane damage induced in cultured human skin fibroblasts by UVA irradiation. Photochem Photobiol 1993; 58(4):510-520.
58. Horio T, Miyauchi H, Sindhvananda J et al. The effect of ultraviolet (UVB and PUVA) radiation on the expression of epidermal keratins. Br J Dermatol 1993; 128:10-15.
59. Nelson D, Gay RJ. Effects of UV irradiation on a living skin equivalent. Photochem Photobiol 1993; 75:830-837.

60. Djavaheri-Mergny M, Maziere C, Dubertret L et al. Ultraviolet A decreases epidermal growth factor (EGF) processing in cultured human fibroblasts and keratinocytes: Inhibition of EGF-induced diacylglycerol formation. J Invest Dermatol 1994; 102:192-196.
61. Braverman IM, Fonferko E. Studies in cutaneous aging. II. The microvasculature. J Invest Dermatol 1982; 78:444-448.
62. Kligman AM. Perspectives and problems in cutaneous gerontology. J Invest Dermatol 1979; 73:39-46.
63. Vermeer M, Schmieder GJ, Yoshikawa T et al. Effects of ultraviolet B light on cutaneous immune responses of humans with deeply pigmented skin. J Invest Dermatol 1991; 97:729-734.
64. Shindo Y, Witt E, Packer L. Antioxidant defense mechanisms in murine epidermis and dermis and their responses to ultraviolet light. J Invest Dermatol 1993; 100:260-265.
65. Urbach F. Incidence of nonmelanoma skin cancer. Dermatol Clin 1991; 9(4):751-755.
66. Blauvelt A, Asada H, Klaus-Kovtun V et al. Interleukin-15 mRNA is expressed by human keratinocytes, Langerhans cells, and blood-derived dendritic cells and is downregulated by ultraviolet B radiation. J Invest Dermatol 1996; 106:1047-1052.
67. Dissanayake NS, Greenoak GE, Mason RS. Effects of ultraviolet irradiation on human skin-derived epidermal cells in vitro. J Cell Physiol 1993; 157:119-127.
68. van Praag MCG, Out-Luyting C, Claas FHJ et al. Effect of topical sunscreens on the UV-radiation-induced suppression of the alloactivating capacity in human skin in vivo. J Invest Dermatol 1991; 97:629-633.
69. Fitzpatrick TB, Szabo S, Wick MM. Biochemistry and physiology of melanin pigmentation. In: Goldsmith LA, ed. Biochemistry and Physiology of the Skin. New York: Oxford University Press, 1983:687-712.
70. Roméro C, Aberdam E, Larnier C et al. Retinoic acid as modulator of UVB-induced melanocyte differentiation. Involvement of the melanogenic enzymes expression. J Cell Sci 1994; 107:1095-1103.
71. Nakazawa K, Nakasawa H, Bonnard M et al. Ca^{2+} and UVB radiation have no effect on E-cadherin-mediated melanocyte-keratinocyte adhesion. Pigment Cell Res 1995; 8:255-262.
72. Archambault M, Yaar M, Gilchrest BA. Keratinocytes and fibroblasts in a human skin equivalent model enhance melanocyte survival and melanin sythesis after ultraviolet irradiation. J Invest Dermatol 1995; 104:859-867.
73. Ho KKL, Halliday GM, Barnetson RStC. Topical and oral retinoids protect Langerhans' cells and epidermal Thy-1^+ dendritic cells from being depleted by ultraviolet radiation. Immunology 1991; 74:425-431.

74. Spencer MJ, Vestey JP, Tidman MJ et al. Major histocompatibility class II antigen expression on the surface of epidermal cells from normal and ultraviolet B irradiated subjects. J Invest Dermatol 1993; 100:16-22.
75. Alcalay J, Craig JN, Kripke ML. Alterations in Langerhans cells and Thy-1+ dendritic epidermal cells in murine epidermis during the evolution of ultraviolet radiation-induced skin cancers. Cancer Res 1989; 49:4591-4596.
76. Hammerberg C, Duraiswamy N, Cooper KD. Reversal of immunosuppression inducible through ultraviolet-exposed skin by in vivo anti-CD11b treatment. J Immunol 1996; 157(12):5254-5261.
77. Rivas JM, Ulrich SE. The role of IL-4, IL-10, and TNF-a in the immune suppression induced by ultraviolet radiation. J Leukocyte Biol 1994; 56:769-775.
78. Simon JC, Cruz Jr PD, Bergstresser PR et al. Low dose ultraviolet B-irradiated Langerhans cells preferentially activate $CD4^+$ cells of the T helper 2 subset. J Immunol 1990; 145:2087-2091.
79. Baadsgaard O, Fox DA, Cooper KD. Human epidermal cells from ultraviolet light-exposed skin preferentially activate autoreactive $CD4^+2H4^+$ suppressor induced lymphocytes and $CD8^+$ suppressor/cytotoxic lymphocytes. J Immunol 1988; 140:1738-1744.

CHAPTER 8

Skin Stem Cell Identification and Culture: A Potential Tool for Rapid Epidermal Sheet Production and Grafting

Lucie Germain, Martine Michel, Julie Fradette, Wen Xu, Marie-Josée Godbout and Hui Li

INTRODUCTION TO SKIN STEM CELLS

Stem cells are fascinating. They are a production factory and give rise to the differentiated tissues. During embryonic development, stem cells multiply and differentiate to generate the cellular diversity necessary for the formation of all tissues. The pluripotentiality of embryonic stem cells is very large. Stem cells in the embryo are probably different from those in adults in the sense that some embryonic stem cells are present only transiently and their pluripotentiality is greater than that of adult stem cells which are limited to one compartment, e.g., blood, cutaneous or intestinal epithelial cells. In this chapter, we will concentrate on skin stem cells of post-natal mammals. During the adult life, stem cells are responsible for tissue homeostasis; they proliferate to maintain the number of differentiated cells at a constant level and replace dead cells or cells lost through injury.

Cell populations from different tissues may be classified into three groups according to their turnover rate: (i) static, which does not divide in the adult such as neurones and skeletal muscle cells; (ii) expanding, which may divide but at a decreasing rate with age

Skin Substitute Production by Tissue Engineering: Clinical and Fundamental Applications, edited by Mahmoud Rouabhia.

such as kidney and liver cells and (iii) renewing, which undergo active division throughout life such as hematopoietic, intestinal and epidermal cells. These three groups of cell populations show similar features during their life span, although at different rates of succession: they originate from embryonic-like elements, go through maturation and senescence..[1]

The skin is a tissue with permanently renewing epithelial cell populations giving rise to the epidermis and the hairs. The main epithelial cells of the epidermis, called keratinocytes, are organized into multiple layers and undergo terminal differentiation as they move through the suprabasal layers toward the skin surface. The outermost layer, the stratum corneum, comprises terminally differentiated anucleated keratinocytes responsible for an important function of the skin which is to act as a barrier limiting fluid evaporation and infection. Squamae from the stratum corneum are continually shed and lost in the environment. Similarly, in the hair follicles, keratinocytes from the outer root sheath undergo terminal differentiation to produce the hair fibers. The cell proliferation is not as constant as in the epidermis since hair growth is cyclical.[2-5] The hair cycle is divided in three phases: The cycle begins with a phase of active growth of keratinocytes from the follicle, the anagen, followed by a phase of degeneration of the lowest part of the hair follicle, the catagen. After a resting phase, the telogen, another cycle starts again.

In renewing tissues, the terminally differentiated epithelial cells have a short life span and must be replaced through proliferation of less differentiated cells. Therefore, it was logical to look for stem cells in these tissues. The rapid and continual turnover of skin and intestinal tissues throughout life rests on a cell population with unique characteristics: the stem cells. Although the existence of stem cells in adult tissues was easier to detect in these rapidly renewing tissues, there is now evidence that stem cells are present in every tissue, even in the liver and the nervous system.[6-16]

The stem cells are expected to possess the following properties: they are relatively undifferentiated, both ultrastructurally and biochemically; they retain a high capacity for self-renewal throughout adult life and have a large proliferative potential; they are normally slow-cycling, but can be stimulated to proliferate in response to injuries and to certain growth stimuli; and finally, they are usually found in well-protected, highly vascularized and innervated areas.[1,17-20]

The study of stem cells was complicated by the fact that it was very hard to identify them at the cellular level. However, clinical experience has given some indications of their tissue localization. It is well known from empirical observations in surgery that a histologically well-differentiated epidermis can be regenerated after its complete destruction (second degree burn) or removal (donor site for split-thickness graft) from the remaining portion of hair follicles and glands. The newly formed epidermis then exhibits a functional barrier to fluid evaporation and infection. In this case, the regeneration of interfollicular epidermis results from the migration of keratinocytes of the hair follicles indicating that stem cells are present in the pilosebaceous region.[21-24]

[^{3}H]thymidine labeling followed by autoradiography has been used to determine the location of proliferating cells in the skin. When [^{3}H]thymidine is injected only once, it labels the cells which proliferate rapidly. It would be unlikely that stem cells incorporate [^{3}H]thymidine following a single exposure since they are slow-cycling. In the skin, the rapidly proliferating cells are concentrated in the basal layer of the epidermis, where they represent 50% of the cells, and in the bulb area of hair follicles.[25-27] Despite their high proliferation rate, these rapidly proliferating keratinocytes have been associated with transit-amplifying cells rather than true stem cells because of their finite self-renewal capacity.[17]

The slow-cycling nature of stem cells is one of their most distinguishing features; this property has been used to localize them.[28-30] The method is based on the fact that several [^{3}H]thymidine injections are necessary to label cells dividing infrequently, and that cells with a long cell cycle will retain the radioactive label over long periods of time. Therefore, one month after multiple [^{3}H]thymidine injections, [^{3}H]thymidine-label-retaining cells represent the ones which divided less frequently.[26,28,29] In keratinocytes which divide rapidly the labeling is diluted at each cell division, and the daughter cells go toward the differentiated compartments (outermost layer in the epidermis and innermost layer in the hair follicles).

In the epidermis, slow-cycling cells were localized in the basal layer.[28,29] A model describing the organization of the epidermis, the epidermal proliferative unit (EPU), has been suggested from kinetic studies of mouse dorsal epidermis. This anatomic site is particular since keratinocytes differentiate regularly and form columns of cells.

According to the EPU model, the skin is comprised of fairly independent proliferative units containing one slow-cycling cell (the stem cell) at the center, giving rise to keratinocytes committed to differentiation (transit amplifying cells) but able to divide rapidly to produce the differentiated progeny (terminally differentiated cells).[28,30,31] This model describes the cell turnover of particularly well-organized epidermis (e.g., mouse dorsal epidermis). However, in humans, the histological organization of the epidermal cells is not as regular. The EPU model does not explain the cellular organizational pattern obtained in the epidermis of chimeric mice.[32]

In the pilosebaceous unit, the bulge area, the portion of the outer root sheath localized under the sebaceous gland to which the arrector pili muscle attaches,[33,34] was identified as the exclusive localization of [^{3}H]thymidine-label-retaining cells of the mouse hair follicle. This led to the conclusion that follicular stem cells reside in this region[35,36] and that they are absent in the bulb.

Various attempts have been made to isolate and purify skin stem cells. The separation of mouse and rat cutaneous epithelial cells on density gradients have led to cultures enriched in cells with progenitor properties.[37,38] However, these cultures have a relatively short lifetime compared to that of human keratinocytes, which can be subcultured for 3 to 20 passages.[39-41] These human cutaneous epithelial cell cultures contain keratinocytes with different capacities for self-renewal. They can be divided into three clonal types: holoclone, meroclone, paraclone.[42] The holoclones are the cells with the highest proliferation potential. They form large colonies, whereas the paraclones only form small abortive colonies. The proliferative potential of the meroclones is between the two.[42] According to the growth potential of these clonal types, the holoclones could represent stem cells and meroclones transit amplifying cells.[17,42] The precise histolocalization of holoclone-forming cells in the tissue biopsy remains uncertain.

Until recently, stem cells have been defined mainly by virtue of their functional attributes. However, such identification was complicated since stem cell characteristics may have sometimes been altered during the assay procedures (for a review, see ref. 20). Moreover, some assays, e.g., [^{3}H]thymidine injections, cannot be performed in humans. The characterization of human tissue stem cell populations remains a difficult task. A new strategy to reach this goal consists of developing phenotypic markers distinguishing progenitors from differentiated cells.[6-16] The identification of an element conserved between species such as structural proteins could permit the characterization of target cells in mouse using well-documented methods before undertaking their characterization in human tissue.

In this chapter, we will describe the identification of keratin 19 as a molecular marker for cutaneous epithelial stem cells. Keratin 19 will then be used to address a number of important issues regarding this population. Does the localization of stem cells vary in different anatomic sites? Does the proportion of stem cells change after birth and in culture? Finally, we will elaborate on the potential implications for the treatment of burn patients, ensuing from stem cell study.

IDENTIFICATION OF KERATIN 19 AS A MARKER FOR SKIN STEM CELLS

We initiated studies to evaluate whether keratin 19 (K19) could be used as a phenotypic marker for skin stem cells. The choice of K19 was based on the fact that this keratin is present in keratinocytes of the bulge area of the human hair follicle[43-45] and that slow-cycling cells have been identified in the bulge area of the mouse hair follicle.[35,36]

Keratins are the main proteins of skin epithelial cells. They assemble into 10 nm filaments that are part of the cellular cytoskeleton.[46] There are over 20 distinct keratins expressed in soft epithelial tissues; many of which are present in various skin epithelial cells.[47] The keratin distribution is specific for each epithelial tissue and varies according to the differentiation status of the cells.[48-51] In the epidermis, the basal cells express mainly keratins 5 and 14 whereas the suprabasal cells, which are more differentiated, express keratins 1 and 10.[52] In the hair follicle, the keratin distribution pattern is more complicated, reflecting the structural complexity of this skin appendage. Beside the above-mentioned epidermal keratins, other keratins, K6, K16 and K17, are highly expressed in certain regions of the hair follicle.[53-56] Some keratins, such as K18, K19 and K20, are only present in a small proportion of cutaneous cells; they are minor keratins in the skin. Hardly detectable biochemically, the minor keratins can be visualized by histolocalization. The K19, a protein of 40 kDa, is the smallest member of the keratin family and has a wide tissue distribution.[43,47,55,57,58] In fetal human skin, K19 is observed in cells of the basal layer.[59,60] In contrast, in adult human skin, K19 is restricted to the outer root sheath of the hair follicle and has not been detected in the epidermis.[43]

In order to assess the possibility that K19 could be a biochemical marker for skin stem cells, we used the well established mouse system to identify [^{3}H]thymidine-label-retaining cells in combination with K19 detection in a double labeling procedure.

The first step was to get antibodies reacting with mouse K19. An antiserum was produced against 14 amino acid residues of the C-terminal sequence of mouse K19.[61-63] To verify the antiserum specificity, total proteins of KLN 205,[64] a cell line derived from mouse squamous cell carcinoma, known to express K19, were separated by SDS-PAGE and then subjected to immunoblot analysis. Figure 8.1A shows that the antiserum reacted specifically to a 40 kDa protein. Figure 8.1B

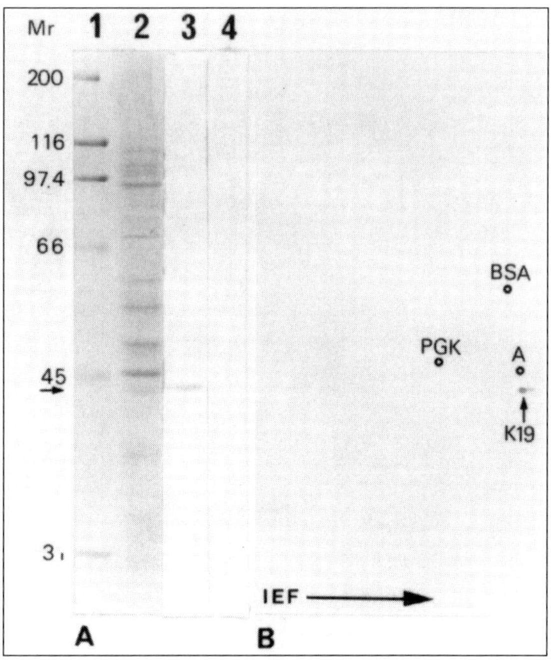

Fig. 8.1. Specificity of guinea pig polyclonal anti-mouse K19 antiserum. Coomassie blue staining of SDS-PAGE (A, lane 2) and western blotting of KLN 205 cell lysate separated by one dimensional (A, lanes 3 and 4) or two-dimensional (B) gel electrophoresis.[100-103] In A, lane 1 is Coomassie blue staining of molecular weight standards separated by SDS-PAGE. Corresponding molecular weights are indicated on the left. Small arrow indicates the position of 40 kDa. Nitrocellulose papers were incubated with guinea pig anti-mouse K19 (A, lane 3; B) or pre-immune (A, lane 4) antisera. In B, the position of protein standards BSA, actin (A) and 3-phosphoglycerokinase (PGK) determined by co-electrophoresis in parallel, and Coomassie blue staining, is indicated. IEF, direction of isoelectric focusing. Note that this antiserum recognizes specifically the mouse K19. Reprinted with permission from Michel M et al, J Cell Sci 1996; 109:1017-1028. ©1996 The Company of Biologists Limited.

establishes that this 40 kDa antigen behaved like an acidic protein, corresponding to K19 according to the numerical classification of Moll et al.[47]

Immunohistological staining of frozen sections of adult mouse skin from various body sites (back, abdomen, muzzle with vibrissae) with this site-directed polyclonal antibody specific to mouse K19 indicated that K19 was present exclusively in the outermost layer of a defined zone of the outer root sheath of the hair follicle, located below the sebaceous gland and corresponding to the bulge area. The lowest part of the hair follicle, the bulb area, and the interfollicular epidermis were not labeled (Fig. 8.2a). The number of K19-expressing cells was lower in smaller hair follicles compared to larger ones (back skin versus vibrissae). This K19 distribution in mouse skin corresponds to that reported in human skin with various antibodies.[43,45,65,66]

The second step was to evaluate if K19 positive cells were slow-cycling—the most distinguishing feature of stem cells. We took advantage of the method of Cotsarelis et al[35] to label mouse slow-cycling cells. [^{3}H]Thymidine was injected subcutaneously twice daily during the first week of postnatal life, a period of synchronized hair growth.[5,67] The label was chased from mice during four weeks and biopsies from back skin were frozen and processed for the double labeling of K19 (by immunofluorescence) and [^{3}H]thymidine (by autoradiography).[61] On transverse sections through the bulge area, the K19-expressing cells were easily detected (Fig. 8.2c,e,g), and the double labeling revealed that they were [^{3}H]thymidine-label-retaining cells (Fig. 8.2d,f,h). Keratinocytes from the bulb region were not labeled with grains nor fluorescence. Therefore, there is a strong correlation between K19-labeling and label-retaining cells into the hair follicle. These results are consistent with the hypothesis that K19 could be a marker of the hair follicle stem cells.

Some [^{3}H]thymidine-label-retaining cells were also observed in the interfollicular epidermis of hairy skin. However, the double labeling experiment revealed that they did not express K19 (data not shown). We interpret this finding as another evidence that stem cells might be absent at this location (see below for discussion). Nevertheless, we undertook experiments to evaluate whether K19 positive cells were present at hairless anatomic sites. Indeed, if K19 is a marker for

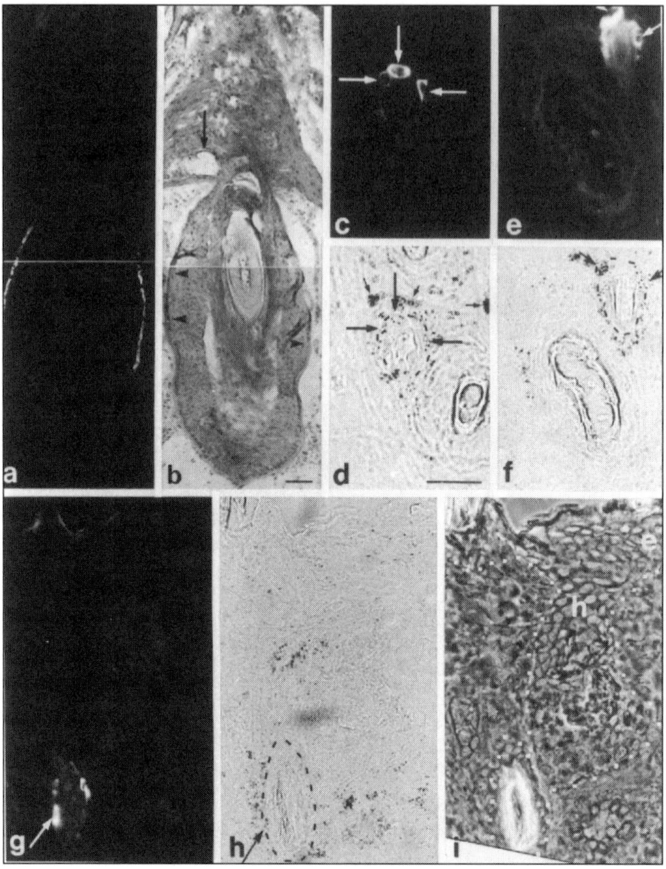

Fig. 8.2. K19-expressing cells of the bulge area are [3H]thymidine-label-retaining cells. Four µm sections were prepared from freshly frozen mouse skin at hairy anatomic sites: mouse vibrissae (oblique section; a,b) or back skin (transverse section; c-i) and processed for immunofluorescence labeling of mouse K19 (a,c,e,g) and autoradiography (d,f,h). (b) Hematoxylin, phloxine and saffron staining of a section consecutive to a. (d,f,h) Bright field micrographs of c,e and g, respectively. (i) Phase contrast micrograph of g. Note that K19-expressing cells are localized in a specific region of the outer root sheath (arrowheads) located under the sebaceous gland (arrow) and identified as the bulge area. They are absent from the interfollicular epidermis. (c and d, e and f, g and h) are double localization of K19-expressing cells (by immunofluorescence) and [^{3}H]thymidine-label-retaining cells (by autoradiography).[61] Note that K19-expressing cells (c,e,g, arrows) also retained [^{3}H]thymidine-labeling (d,f,h, arrows). The black dotted lines in d,f and h indicate the junction between the epithelium and the mesenchyme surrounding hair follicles. Small arrows indicate mesenchymal cells retaining [^{3}H]thymidine-labeling (d). The number of K19 positive cells varies with hair size, being lower in small hair follicles of back skin. The white dotted line in i represents the dermo-epidermal junction. h: hair; e: epidermis. Bars: (a,b) 50 µm; (c-i) 25 µm. Reprinted with permission from Michel M et al, J Cell Sci 1996; 109:1017-1028. ©1996 The Company of Biologists Limited.

cutaneous epithelial stem cells, it must label those present in glabrous anatomic sites. These stem cells have been identified in the epidermis, at the bottom of the deep rete ridges on the basis of ultrastructural characteristics.[19,68]

DIFFERENTIAL LOCALIZATION OF STEM CELLS AT HAIRY AND GLABROUS SKIN SITES

Our interest in the treatment of burn patients led us to continue our studies on human tissues. Adult human skin samples from various anatomic sites were labeled with Ks19.1, a mouse monoclonal anti-human K19 antibody, in order to determine if K19-expressing cells were always restricted to the bulge area of the hair follicle. At all hairy skin sites studied, K19-expressing cells were localized in a well-defined zone of the hair follicle below the sebaceous gland, the bulge area (Fig. 8.3a). This is the deepest region of the permanent portion of the hair follicle. K19-expressing cells were detected throughout the hair cycle, in telogen follicles (the resting phase that follows the degeneration phase; Fig. 8.3a) as well as in anagen follicles (the growing phase of hair follicles; Fig. 8.3a). The number of cells expressing K19 varied with hair size. A greater number of K19-expressing cells were detected in the large hair follicles of scalp compared to small follicles from breast or abdomen. Thus, these data are consistent with the hypothesis that K19 is a marker for skin stem cells since K19 positive cells are located in the deepest permanent portion of the hair follicle and do not degenerate during the hair cycle, as expected from a stem cell population.

In the skin, two cell types express keratins: keratinocytes, the main epithelial cell type comprising 95% of the epidermal cells, and Merkel cells, a minor cell population representing less than 1% of the epidermal cells. Merkel cells express the intermediate-sized filaments of epithelial cells, specifically K8, K18, K19, and K20 but they also contain secretory granules typical of neurones.[69-73] Their precise origin and role remain to be elucidated. However, if they are epithelial cells, it is likely that they originate from the same stem cells as do keratinocytes. According to this hypothesis, the cutaneous epithelial stem cells would be pluripotential, giving rise to Merkel cells and keratinocytes.

We have confirmed that the K19-expressing cells present in the bulge area of the hair follicle were not Merkel cells by

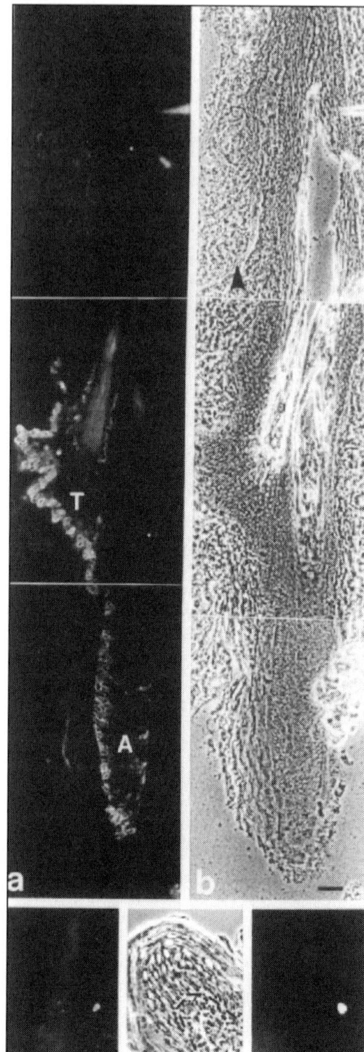

Fig. 8.3. K19 and K18 expression in hairy anatomic sites of human skin. Indirect immunostaining of K19 (a,c) and K18 (e) of breast skin. (b,d) Phase contrast micrographs of a and e, respectively. (a) Note that K19-expressing cells are present in a specific region of the outer root sheath under the sebaceous gland (arrow in b). This longitudinal section of a new hair shows that K19-expressing cells are present in anagen (A) as well as telogen (T) follicles. The few K19-expressing cells present in the basal layer of the epidermis are also K18-positive indicating that they are Merkel cells. Bars: (a,b) 50 μm; (c-e) 100 μm. Reprinted with permission from Michel M et al, J Cell Sci 1996; 109:1017-1028. ©1996 The Company of Biologists Limited.

immunolocalizing K18 (using Ks18.174 antibody)[61,71] and K19 on the same section of human skin (Fig. 8.4). Merkel cells were present in the area above the bulge, the isthmus, while the bulge region was mainly negative for K18 (Fig. 8.4). A similar distribution of Merkel cells was reported by Moll[70] using an anti-K20 antibody. In contrast, the rare K19 positive cells detected in the interfollicular epidermis also coexpressed K18, indicating that they were all Merkel cells (Figs. 8.3c,e, 8.4a,c). Therefore, we have obtained similar results on the localization of K19 positive cells for mouse and human skin at hairy anatomic sites.

We then decided to examine glabrous skin where stem cells have been identified in the epidermis, at the bottom of the deep rete ridges.[19] In human glabrous skin, small K19 positive cells were present in the

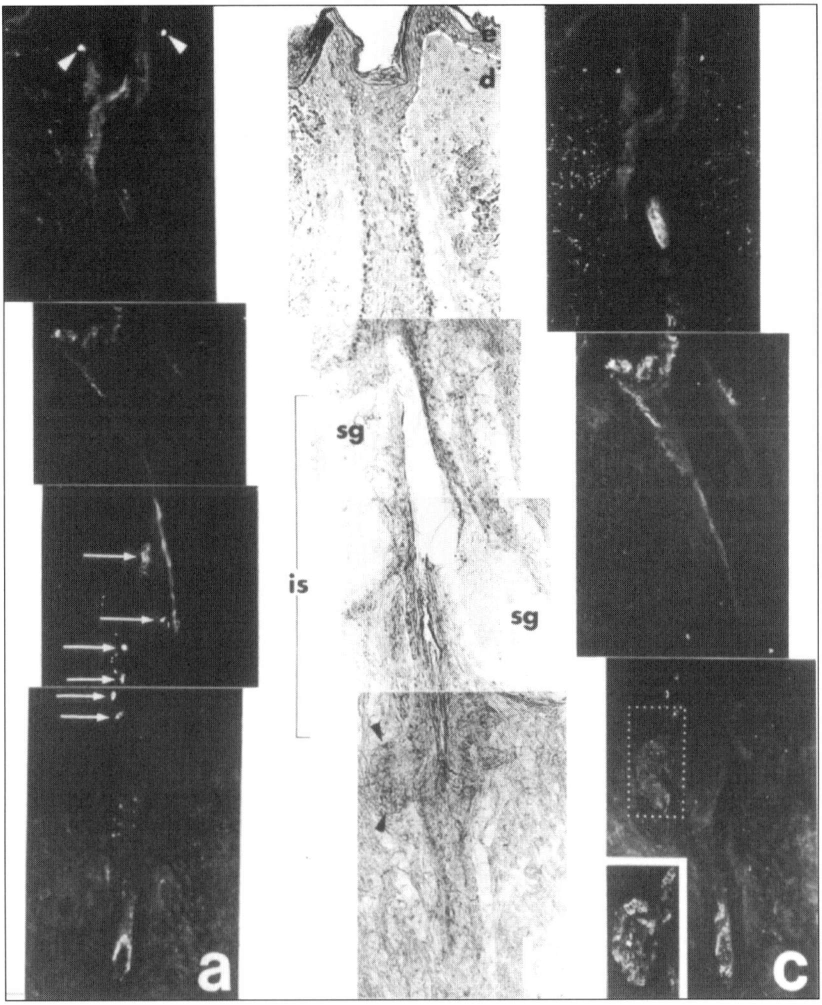

Fig. 8.4. Double-labeling of a longitudinal section of a human breast skin hair follicle with anti-K18 (a) and anti-K19 (c) antibodies. (b) Hematoxylin-phloxine-saffron staining of a consecutive section. The insert in c represents K19 labeling of the bulge area (dotted rectangle), taken from a consecutive section at twice the magnification. Note that Merkel cells, as identified by K18, are present in the basal layer of the interfollicular epidermis (white arrowheads in a). In a, most of K18-positive cells (white arrows) were localized above the bulge area, in the isthmus region of the hair follicle. Most of them were also K19-positive (c). The bulge area indicated in b by black arrowheads contained cells that were positive to K19 (c) but clearly not to K18 (a). The dashed line in b indicates the dermo-epidermal junction. e, epidermis; d, dermis; sg, sebaceous gland; is, isthmus. x 130. Reprinted with permission from Fradette J et al, Biochem Cell Biol 1995; 73:635-639.

basal layer of palm (Fig. 8.5a) and sole epidermis. They were localized in the deep part of the rete ridges (Fig. 8.5a,b). These K19 positive cells did not contain K18 (Fig. 8.5c), indicating that they were not Merkel cells. Merkel cells coexpressing K18 and K19 were also frequently observed.[71] At hairless anatomic sites of mouse skin, sole epidermis, the localization of K19-expressing cells revealed that they were localized at the bottom of the rete ridges (Fig. 8.6). Thus, in contrast to hairy skin, a population of cells expressing K19 but not

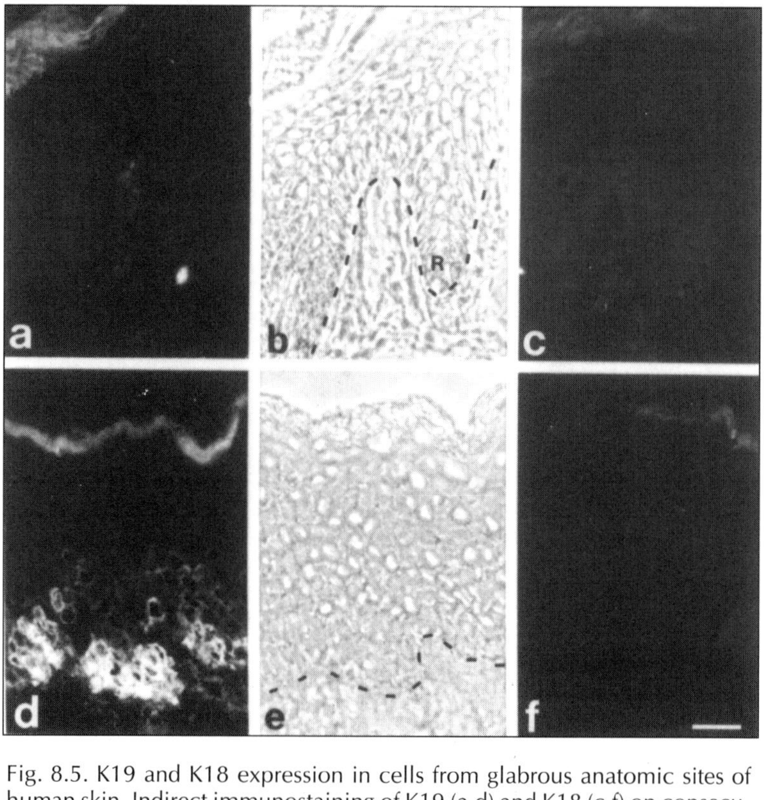

Fig. 8.5. K19 and K18 expression in cells from glabrous anatomic sites of human skin. Indirect immunostaining of K19 (a,d) and K18 (c,f) on consecutive four mm thick sections of finger skin of a one-year-old child (a-c) and from foreskin of a 2-day-old child (d-f). (b and e) Phase contrast micrographs of c and f, respectively. Note the presence of a small K19 positive cell, that does not contain K18 (c), in the deep part of the rete ridge (R; a). In newborn foreskin, the expression of K19 is present in most of the basal cells (d), the absence of K18 labeling (f) indicates that they are not Merkel cells. The dermo-epidermal junction is indicated by dashed line. Bar: 25 µm. Reprinted with permission from Michel M et al, J Cell Sci 1996; 109:1017-1028. ©1996 The Company of Biologists Limited.

K18 was observed in the epidermis of glabrous skin at a location similar to that expected for stem cells.[19,68]

In light of these data, two hypotheses can be generated. First, cutaneous epithelial cells are present in the epidermis at hairy as well as glabrous anatomic sites. This would imply that K19 labels only part of the stem cells since it does not label the slow-cycling cells of the interfollicular epidermis at hairy skin sites. Second, the location of cutaneous epithelial stem cells varies according to anatomic site. In glabrous skin, they are present in deep rete ridges of the epidermis, whereas in hairy skin they are restricted to the bulge area or slightly lower portion of the hair follicle and absent from the interfollicular epidermis. In this case, K19 could label all the cutaneous epithelial stem cells. This would also imply that the [^{3}H]thymidine-label-retaining cells detected in the interfollicular epidermis are not stem cells. Other evidence suggests that this could be the case. Indeed, recent in vitro studies comparing the growth potential of various subpopulations of follicular keratinocytes to that of interfollicular epidermis have shown that cells isolated from the bulge area or slightly lower in the hair follicles had a significantly greater proliferative potential and were more clonogenic than cells from the bulb[74] or interfollicular epidermis.[75] These results lend strong support to the idea that skin stem cells are located in the bulge area or slightly lower in the hair follicle

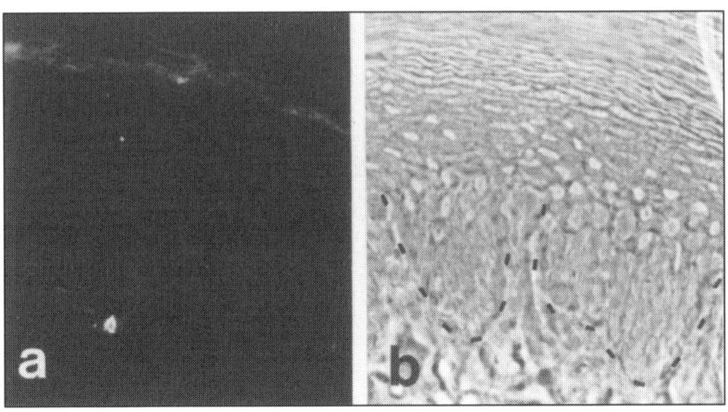

Fig. 8.6. K19 expression in cells from glabrous anatomic sites of mouse skin. Indirect immunostaining of K19 (a) on 4 μm thick section of plantar mouse skin. (b) Phase contrast micrograph of a. The dermo-epidermal junction is indicated by dashed line. Note the presence of a small K19 positive cell in the deep portion of the rete ridge indicating that K19-expressing cells are present in the epidermis at this anatomic site.

and that they are absent in the interfollicular epidermis. Furthermore, a characteristic usually associated with stem cells, their location in a well-protected site, is lacking from the interfollicular epidermis since the tips of the papilla can be easily removed by abrasion. Indeed, the thickness of the interfollicular epidermis varies with anatomic sites but it is always thin compared to sole or palm epidermis. For example, in the mouse, the interfollicular epidermis of the cheek is 4 to 6 layers thick (Fig. 8.2b) compared to more than 20 layers in the sole epidermis (Fig. 8.6b). In humans, the interfollicular epidermis is thicker than in mice (Fig. 8.3d compared to Fig. 8.2b) but the epidermal papilla goes much deeper in the dermis at glabrous (sole and palm) than hairy skin sites. Therefore, cells located at the bottom of the epidermal papilla are better protected in glabrous than hairy anatomic sites. Nonetheless, additional studies will be necessary to elucidate the question whether the interfollicular epidermal slow-cycling cells are true stem cells or if they could rather be regarded as amplifying cells as suggested by Lane et al.[44] Thus, because this issue has yet to be resolved, it is difficult to determine whether K19 is expressed in all cutaneous stem cells.

In skin, stem cells are known to be within the cell population expressing high levels of $\alpha_2\beta_1$ and $\alpha_3\beta_1$ integrins.[76,77] To determine if K19-expressing cells were in the proportion of keratinocytes expressing high levels of β_1 integrins, we have performed double labeling of K19 and α_3 integrin subunit (VM2 monoclonal antibody).[78] At hairy skin sites, K19-expressing cells (Fig. 8.7c,e) of the bulge area of the outer root sheath were also integrin-bright cells (Fig. 8.7d,f). Similarly, in glabrous tissue, few small K19 positive cells (Fig. 8.7a) were also integrin-bright cells (Fig. 8.7b). It is interesting to note that they were among the brightest α_3 integrin positive cells of the epidermis (Fig. 8.7b). However, K19 positive cells represent only a proportion of the α_3 integrin positive cells. The β1 integrin-bright cells encompass a much larger compartment than the K19-expressing cells, based on the higher percentage of cells expressing high level of β1 integrins (e.g., 25% bright cells in the basal layer of palm)[77] compared to K19 (Fig. 8.7). Our results are consistent with the fact that K19 positive cells are a subset of the $\alpha_3\beta_1$ integrin-bright cell population, which is otherwise too important to be uniquely specific to stem cells. We suggest a model of cutaneous epithelial cell differentiation in which K19-expressing stem cells give rise to K19-negative β1 integrin-bright cells

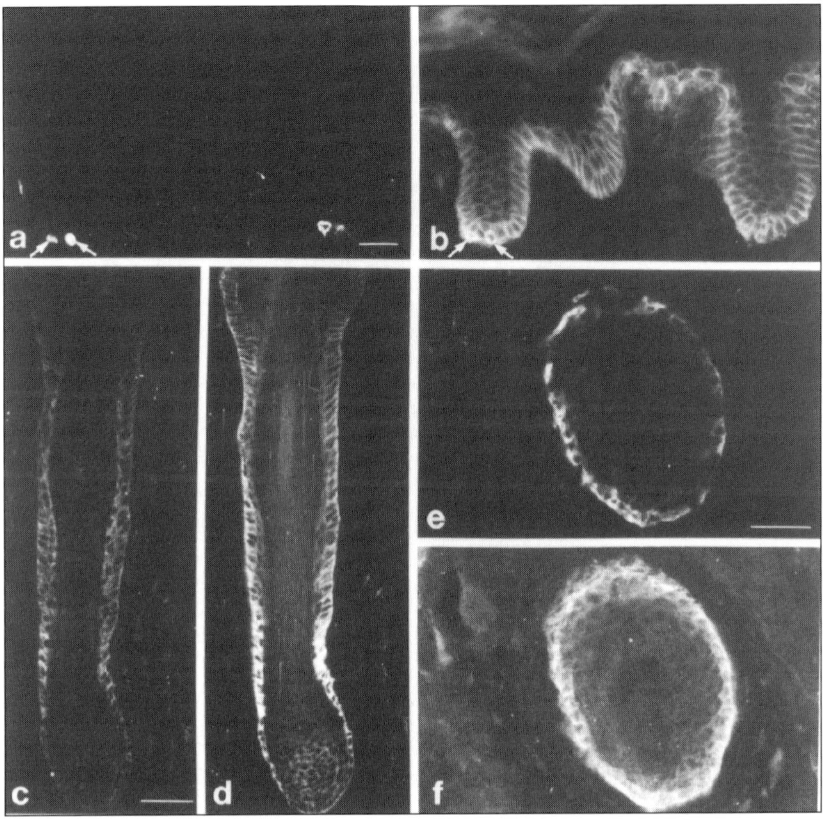

Fig. 8.7. K19 and $\alpha_3\beta_1$ integrin expression in cells from hairy and glabrous anatomic sites of human skin. Epidermis from palm skin (a,b) or hair follicle from breast skin (c,d,e,f) were double labeled with an antibody to K19 visualized with rhodamine-conjugated second antibody (a,c,e) and an FITC-conjugated α_3 integrin subunit (b,d,f). Note that in glabrous skin (a,b), small K19 positive cells of the deep part of the rete ridge are also integrin-bright cells (arrows), another K19 positive cell, which is not integrin bright, is not round and probably represents a Merkel cell. In hairy skin, the K19-expressing cells are also integrin-bright cells. Bars: (a,b,e,f) 25 μm, (c,d) 50 μm. Reprinted with permission from Michel M et al, J Cell Sci 1996; 109:1017-1028. ©1996 The Company of Biologists Limited.

which differentiate in transitional cells and finally in differentiated keratinocytes (Fig. 8.8).

Therefore, the following evidence supports the hypothesis that K19 could constitute a phenotypic marker for skin stem cells. First, K19-expressing cells of the bulge area of hair follicles are slow-cycling cells. Second, our identification of K19-expressing cells at the bottom of deep rete ridges in glabrous skin (palm and sole) is consistent with

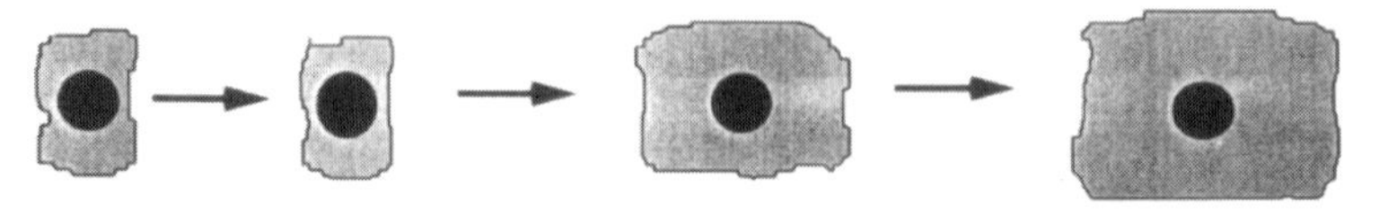

Fig. 8.8. Schematic model of cutaneous epithelial cell differentiation.

previous work identifying stem cells in this region based on their morphological and functional characteristics.[19,68] Third, the expression of high levels of the $\alpha_3\beta_1$ integrin at the surface of K19 positive cells present at hairy and glabrous skin sites is consistent with the previous demonstration that skin stem cells express higher amounts of β_1 integrin than differentiated keratinocytes.[76,77] If the hypothesis that K19 labels all cutaneous stem cells is true, our results imply a differential stem cell localization depending on the anatomic site.

In the following sections, we will take advantage of the K19 labeling method to address important issues regarding the skin stem cell population. We will exploit the concept that K19 could be a marker for cutaneous stem cells to perform in situ and in vitro studies. Since K19 is not restricted to putative cutaneous stem cells, Merkel cells also express K19, we have chosen foreskin where the Merkel cell density is relatively low for the following experiments. (ref. 79, Fradette J. and Germain L.; unpublished observations).

SKIN STEM CELL PROPORTION DECREASES AFTER BIRTH

Whether the number of stem cells varies during post-natal life remains a matter of debate. In order to evaluate age-associated changes in the number of human skin stem cells, we analyzed cell populations from the same body site, foreskin, in newborns, 2-5- and 17-42-year-old individuals. To evaluate if the distribution of K19-expressing cells varied with the age of the person, immunofluorescence labeling was performed on foreskin from 2-day- to 42-year-old donors. K19-expressing cells were present in the basal layer of the epidermis. They were more numerous in newborns (Fig. 8.9a) than in children (Fig. 8.9b) and, again, than in adults (Fig. 8.9c). The proportion of positive cells present in the basal layer was quantified on tissue sections. In neonatal foreskin (2-7-day-old boys), K19 expression was observed in most of the cells of the epidermal basal layer (~85%; Figs. 8.5d, 8.9a). These cells did not contain K18 (Fig. 8.5f). In the epidermis of older donors the K19-expressing cells were restricted to the basal layer in the shallow region of the papilla. They represented ~10% and ~0.1% of the basal layer in 2-5-year-old children and 17-42-year-old men, respectively. K19 positive cells were also present in the few hair follicles observed in these two age groups whereas foreskins from 2-7-day-old boys were glabrous.

To better quantify the age-related variations in the proportion of K19-expressing cells, epithelial cells were isolated from newborn, child and adult foreskins by a double digestion method with thermolysin and trypsin (Fig. 8.10).[80] These single cell populations were analyzed by flow cytometry after K19 labeling.[61] The results show an age-dependent decrease of the proportion of K19 positive cells in foreskin. In adults, less than 1% of the cells expressed K19 (Table 8.1).

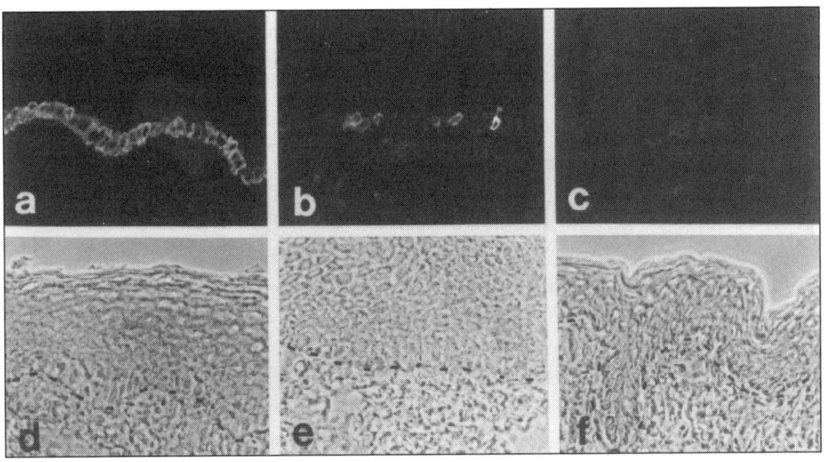

Fig. 8.9. K19 expression in newborn, child and adult human foreskins. Indirect immunostaining of K19 (a-c) on four μm thick sections of foreskin of a newborn (a) child (b) and adult (c). (d-f) Phase contrast micrographs of a-c, respectively. Note the diminution in the number of K19-expressing cells as the age of the donor increases. Some fields of adult foreskin are negative (c). The dermo-epidermal junction is indicated by a dashed line.

Our study clearly established that the number of K19-expressing cells decreased after birth. The two methods used to quantitate the proportion of K19-expressing cells in situ on tissue sections, and in vitro after cell dissociation, led to the same conclusion: the proportion of K19-expressing cells diminishes from neonates to children to

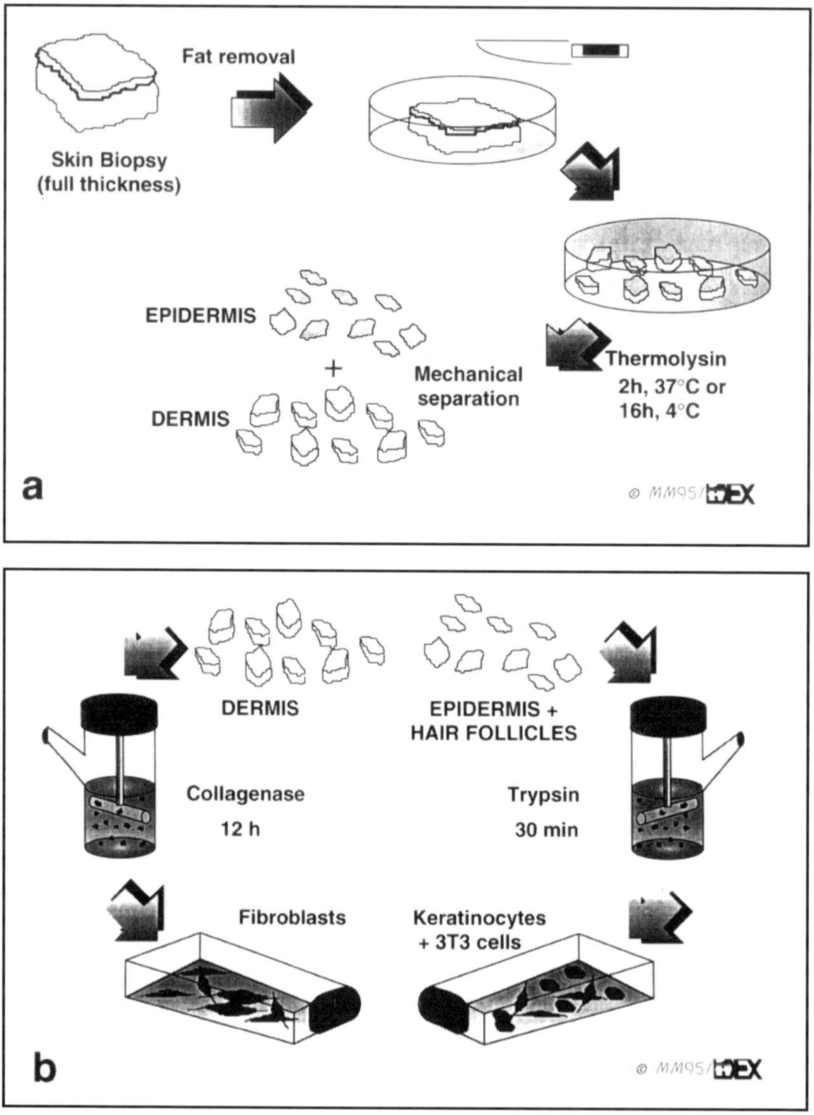

Fig. 8.10. Schematic representation of the thermolysin and trypsin method for the differential isolation of epithelial cells and fibroblasts from human skin biopsies.

adults. Based on the hypothesis that K19 labels stem cells, this would indicate a decrease in the number of stem cells after birth.

SKIN STEM CELL PROPORTION VARIES DURING MONOLAYER CULTURE

Human keratinocytes can be cultured in vitro. They can undergo several passages in culture.[41] However, these cultures are heterogeneous. They contain three clonal types present in different proportions in newborn and in adult cell cultures.[42] The culture lifetime also varies according to donor age. In order to evaluate the fate of K19-expressing cells upon keratinocyte culture, epidermal cells from neonatal human foreskin were stained with the anti-K19 antibody. One

Table 8.1. Percentage of K19 expressing cells in function of the anatomic site, age, and culture time*

	Anatomic Sites			
	Foreskins			Breast
Passage	2-7 days	2-10 years	17-42 years	16-64 years
Fresh cells	7.5 (3.5) ‡ (n=3)	1.5 (1.7) (n=2)	0.2 (0.2) (n=5)	0.7 (1.0) (n=4)
P0	21.3 (12.9) § (n=10)	5.4 (5.7) (n=5)	1.7 (1.8) (n=5)	1.2 (0.9) (n=4)
P1	24.8 (13.0) (n=10)	9.3 (5.3) (n=4)	1.7 (1.5) (n=6)	1.9 (1.7) (n=4)
P4-P5	17.1 (9.4) (n=6)	10.3 (5.8) (n=2)	1.9 (0.4) (n=2)	1.2 (1.0) (n=4)
P6	11.8 (7.9) (n–2)	ND	ND	ND
P7-P8	3.6 (5.5) (n=4)	ND	0.6 (0.6) (n=3)	ND

* Cell suspensions were fixed and analyzed by FACS®. The percentage of K19 positive cells was determined in a FSC gated (0-200) cell population as described in Materials and Methods.

‡ Cells were isolated by the double digestion method with thermolysin and trypsin, fixed and stained as described in Materials and Methods. Mean (s.d.).

§ For primary culture (P0) or serial subcultures (P1 to P8), cells were trypsinized at 70-80% of confluence, fixed and stained as described in Ref. 61. Reprinted with the permission from Michel et al, J Cell Sci 1996; 109:1017-1028. © 1996 The Company of Biologists Limited.

day after plating freshly isolated epidermal cells, most of the attached keratinocytes were still round. The K19-expressing cells were found primarily as single cells (Fig. 8.11a). After 2 days, keratinocytes had spread onto the surface. Several K19 positive cells were arranged in pairs. The characteristic filamentous pattern was detected in each cell of the doublet (Fig. 8.11b). After one week in culture, numerous colonies were formed. Most of them contained several positive cells. K19-expressing cells were mainly observed in small keratinocytes located at the periphery of the colonies. In the following subcultures, similar patterns of K19 expression were detected, cells harboring filaments being at the border of the colonies (Figs. 8.11c, 8.12a). These results suggest that K19-expressing cells proliferate in culture. Moreover, K19-expressing cells were present at the periphery of several large colonies which is consistent with the finding that proliferating keratinocytes are localized in the outermost layers of the colonies.[81]

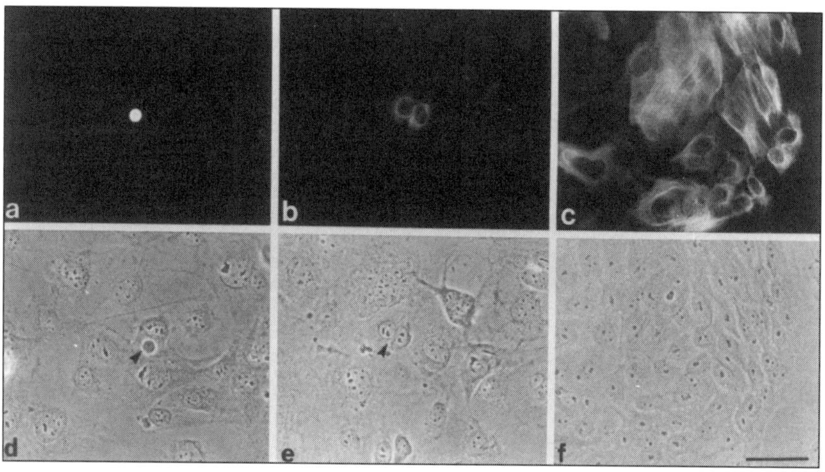

Fig. 8.11. Detection of K19-expressing cells in human keratinocyte culture. Indirect immunostaining of K19 on human keratinocytes, isolated from foreskin of a 2-day-old-child, and cultured in vitro (a-c). (d-f) Phase contrast micrographs of a-c, respectively. Keratinocytes (1 x 10^4 /cm^2) and 3T3 (6 x 10^4 /cm^2) feeder layer were seeded on glass slides. (a,b,d,e) represent freshly isolated keratinocytes, (c,f) are second subculture of keratinocytes. Note that 24 hours after the seeding of freshly isolated cells, the K19 is expressed in round cells (a). After 2 days in culture, cells are spread, and doublets of K19-expressing cells are observed (b), indicating that these cells proliferate. After one subculture, K19-expressing cells are small and mainly located at the periphery of the large colony (c,f). Arrowheads (d,e) point to labeled cells in (a,b) respectively. Bar: 50 μm. Reprinted with permission from Michel M et al, J Cell Sci 1996; 109:1017-1028. ©1996 The Company of Biologists Limited.

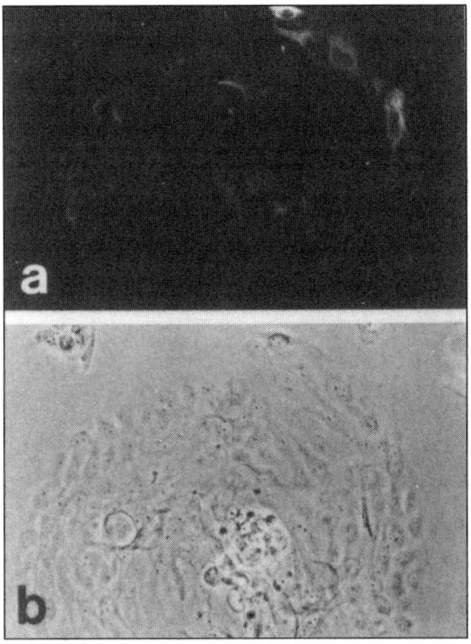

Fig. 8.12. Detection of K19-expressing cells in human keratinocyte culture. Newborn foreskin keratinocytes were cultured and labeled as described in Fig. 8.11. (a) K19 immunostaining. (b) Phase contrast micrograph of a. Note that the K19-expressing cells are located at the periphery of the colony.

To insure that K19-expressing cells present in culture were not Merkel cells, double labeling of K19 and K20 was performed on primary cultures or first subcultures of neonatal foreskin epidermal cells. Single K19-expressing cells, as well as doublets and cells at the periphery of the colonies, did not contain K20, indicating that they were not Merkel cells.

The fate of K19-expressing cells in culture was also evaluated in cell populations isolated from mouse skin. In primary culture of mouse epithelial cells, K19-expressing cells were observed as single cells after 2 days, pairs after 2 to 4 days, and colonies after one week. Therefore, the similar results observed with mouse and human skin clearly establish that the number of K19 positive cells increases during the first subcultures.

The life span of keratinocyte culture decreases as the age of the donors increases (refs. 39, 40, 82, Michel M, Fradette J, Germain L; unpublished observations). To evaluate if the fate of these cultures was related to the number of K19-expressing cells, freshly isolated cells and various passages of newborn, children and adult foreskin keratinocyte cultures were analyzed by flow cytometry with a FACS®. Figure 8.13 shows the fluorescence intensity (FL1) and the forward light scatter (FSC) of newborn keratinocytes (passage 1) stained with an anti-K19 antibody. A population of K19 positive cells (Fig. 8.13b)

Fig. 8.13. Flow cytometry analysis of K19 expression in a population of human keratinocytes extracted from foreskin of a 2-day-old-child. FSC and K19 immunostaining (b) or control (a) are represented as dot plots. After the second subculture, cells were labeled with anti-K19 antibody (b) detected with an FITC-conjugated second antibody and examined with a flow cytometer. FSC and fluorescence were measured in arbitrary units on a linear and logarithmic scale, respectively. In controls, the primary antibody was replaced by PBS (a). The percentage of K19-expressing cells was 35% in the experiment shown. Reprinted with permission from Michel M et al, J Cell Sci 1996; 109: 1017-1028. ©1996 The Company of Biologists Limited.

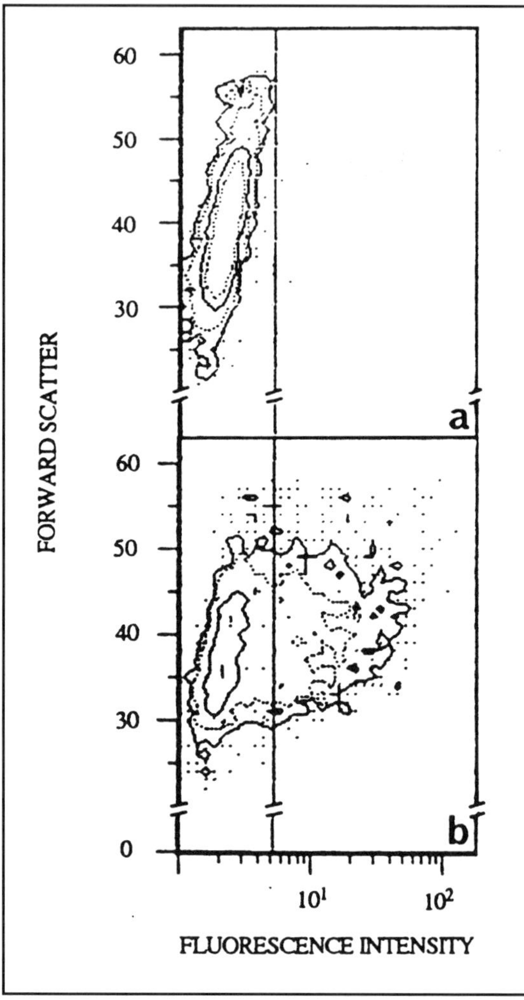

can be easily distinguished from control unlabeled cells (Fig. 8.13a). The labeling of large keratinocytes (identified by FSC values over 200) was considered as nonspecific since differentiated keratinocytes were also variably stained by unrelated antibodies (anti-digoxigenin). Therefore, data were gated for FSC (0 to 200) in further analysis. The percentage of K19 positive cells was determined by subtracting the control from the labeled cell profile and evaluating the proportion of the cells under the peak or shoulder. The results show that the percentage of K19 positive keratinocytes in cell populations freshly isolated from foreskin decreased when the age of the donors increased.

When these epithelial cell populations were cultured, the percentage of K19 positive keratinocytes significantly increased in primary culture, remained high for the first subcultures and decreased before the senescence of the cultures. These variations were observed in all age groups (Table 8.1). Similar percentages of K19 positive cells were obtained in keratinocytes isolated from adult foreskin and breast skin (Table 8.1).

These results show that the proportion of K19 positive keratinocytes diminishes with increasing age and that the number of K19 positive cells varies in culture. It could help to explain the well-documented empirical observation that newborn foreskin keratinocytes can undergo a greater number of passages in culture than adult trunk keratinocytes and that they can generate a greater total number of keratinocytes. The number of passages that newborn cells can undergo in culture is also higher.[39,40,82] The results of our study suggest that the life span of cultured keratinocyte populations depends on the number of stem cells initially seeded in primary culture. Indeed, we observed a correlation between the percentage of K19-expressing cells in freshly isolated keratinocytes and the number of subcultures that could be generated in vitro from these cell populations. Based on the notion that stem cells have, by definition, an extensive capacity for self-renewal in vivo,[17,20] it seems likely that the greater the number of stem cells is, the longer is the lifetime. Therefore, considering the hypothesis that K19 is a marker for stem cells, the number of K19-expressing cells in freshly isolated populations would be a good parameter to predict the lifetime of in vitro cultures.

As mentioned before, keratinocytes in culture can be divided into three clonal types (holoclone, meroclone, paraclone) with different capacities for self-renewal, the holoclones being the cells with the highest proliferation potential.[42] Interestingly, the percentages of K19-expressing cells that we found in cultured keratinocyte populations are similar to the percentages of holoclones generated by these cultures as previously reported by Barrandon and Green.[42] Indeed, newborn foreskin contained 28% holoclones (25% of K19-expressing cells) whereas adult breast skin contained 3.1% holoclones (1.9% of K19-expressing cells). Thus, a lower number of K19 positive cells correspond to a lower number of holoclones and a decline in culture lifetime, as donor age increases. Therefore, we propose that holoclones are generated by K19 positive cells.

If we assume that K19 labels stem cells, our results can be interpreted in terms of the fate of stem cells in culture. The increased frequency of K19-expressing cells in primary culture compared to cells freshly isolated from the epidermis suggests that the proliferation rate of these cells in culture is higher than that of other keratinocytes. It is not surprising that stem cells proliferate in vitro. Indeed, stem cells are usually slow-cycling in vivo but they can be induced to proliferate by stimuli such as wounding. The in vitro culture system is likely another stimuli for the proliferation of stem cells since they are isolated from their neighbors and try to regenerate the epidermis. Our immunofluorescence data (Fig. 8.11) and the slow increase of K19 positive cell number suggest that they proliferate in culture. After several subcultures, the proportion of K19-expressing cells decreased and the number of differentiated cells increased before the senescence of the culture.

POTENTIAL IMPLICATIONS FOR BURN PATIENT TREATMENT

Biopsy Harvesting for Keratinocyte Culture: Choice of Anatomic Site

The results obtained show that the number of K19-expressing cells varies according to hair size. This might have implications on the anatomic site chosen to harvest the biopsy from which the epidermal cells are extracted. Indeed, if K19 is a marker for skin stem cells, the greater number of K19-expressing cells in larger hair follicles suggests that more stem cells could be isolated in skin sites containing large hair follicles such as scalp or armpit. Since biopsies containing more K19-expressing cells give rise to keratinocyte cultures with longer life span in vitro (e.g., newborn compare to adult), it is likely that the anatomic site allowing the best yield in K19-expressing cells should be chosen. Work is in progress to compare the yield and culture behavior of keratinocytes isolated from two different anatomic sites of the same donor in order to evaluate if the theoretical superiority of anatomic sites containing larger hairs can be verified experimentally.

Human Keratinocyte Isolation Method

The results obtained on the differential localization of K19-expressing cells in hairy and glabrous anatomic sites might have important implications for the culture of keratinocytes if we accept that K19 is a marker for skin stem cells. Until now, most methods designed for the extraction of keratinocytes have focused on the epidermis. This is appropriate for glabrous anatomic sites. However, if stem cells are exclusively located in the bulge area of hair follicles at hairy anatomic sites, we must develop methods to optimize the isolation of keratinocytes from hair follicles.

A few years ago, we have described a double digestion method for the extraction of keratinocytes (Fig. 8.10). At that time, we verified that thermolysin separated the dermo-epidermal junction of the interfollicular epidermis.[80] This two-step digestion method with thermolysin and trypsin was advantageous over the conventional trypsin method since the cultures were not contaminated by fibroblasts, the efficiency of keratinocytes to form colonies was increased and a shorter time was necessary for the keratinocytes to reach confluence. We interpreted the better results obtained with this method, compared to the trypsin method[39,83] for the isolation of keratinocytes from breast skin, as an indication that more epidermal basal cells were isolated with the new method. However, in light of the results presented above, the improvement might be due to the isolation of keratinocytes from the bulge area of hair follicles. Therefore, we have evaluated the effect of thermolysin on the separation of hair follicles from the dermis. Figure 8.14 shows that when the epidermis is separated from the dermis with forceps after thermolysin treatment, the hair follicles and glands are also detached from the dermis and can be peeled with the epidermis. Moreover, the K19-expressing cells are still present in the external layer of the outer root sheath (Fig. 8.15). Thus, it is likely that the difference between these two isolation methods lies in the fact that the yield of stem cells (K19-expressing cells) is higher with the thermolysin and trypsin method than with the trypsin method.

Skin Banking

Chapter 5 describes that the grafting of epidermal sheets composed of allogeneic and syngeneic cells leads to the formation of a syngeneic epidermis. To apply this technique to burn patients, one will have to raise banks of human keratinocytes. The best candidates for the production of large banks of keratinocytes are stem cells because of their longer lifetime. Newborn foreskin is probably one of

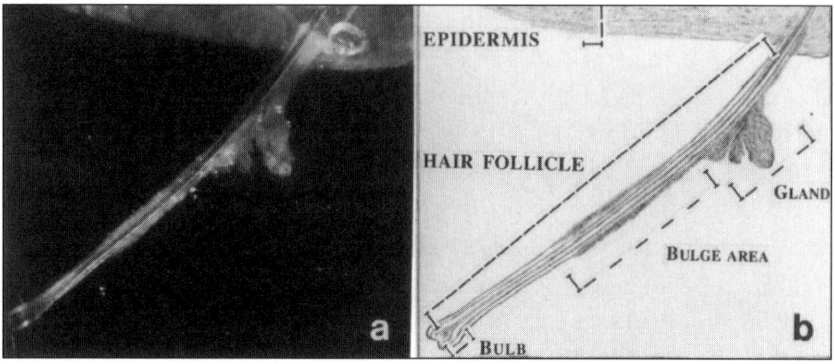

Fig. 8.14. Epithelial portion of human skin obtained after thermolysin digestion. Breast skin was digested with thermolysin overnight at 4°C. The epidermal portion was separated from the dermis with forceps. (a) Photograph taken under binocular of the epithelial portion of the skin. (b) Schematic representation of the epidermis and hair follicle shown in a. Note that the hair follicle and sebaceous gland are detached from the dermis after thermolysin digestion and that they can be peeled with the epidermis.

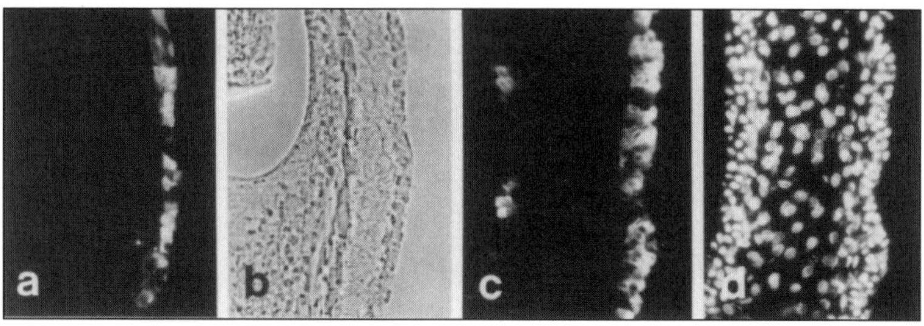

Fig. 8.15. K19 expression in the epithelial portion of human skin obtained after thermolysin digestion. Breast skin was digested with thermolysin overnight at 4°C. The epidermal portion was separated from the dermis with forceps. (a,c) K19 immunostaining of longitudinal section of the hair follicle. (b) Phase contrast micrograph of a. (d) Hoechst staining of c. Note that the K19-expressing cells are present in the epithelial portion of the skin after thermolysin digestion and mechanical separation (a,c).

the best tissues to start with, since it contains the greatest proportion of stem cells. Moreover, newborn foreskin keratinocytes can be kept in culture for a large number of passages.

Furthermore, skin banks may be used in the production of tissue engineered skin. Various models have been proposed to produce a dermo-epidermal equivalent in vitro.[84-92] These equivalents have different applications in vitro (for pharmacological testing) and eventually in vivo (as skin replacement for burn patients). We have shown that skin equivalents can be produced from newborn or adult keratinocytes.[93] In the case of in vitro studies, the production of a large number of similar skin equivalents is highly suitable in order to minimize the data variability due to replicates (e.g., percutaneous absorption data). In this context, the possibility for stem cells to raise large progeny is advantageous for the production of a high number of skin equivalents from the same starting material to reduce the variability between skin equivalents.

Wound Healing Variation as a Function of the Patient's Age

Our results show that the proportion of K19-expressing cells diminishes as the age of the donor increases indicating that the proportion of stem cells decreases after birth. This result is of interest considering the expected role of stem cells in wound healing. It could explain the clinical observations that children heal faster than adults.[94-96] For burn patients, this parameter is clinically relevant since the speed of closure of burn wounds influences the mortality rate.[97,98] One of the methods to predict the probability of survival after burn injury has been proposed by Baux and modified later.[99] The original Baux rule predicts a poor outcome when the age of the patient added to the percentage of the burn is greater than 75. A more accurate method for determining the chance of survival is based on the "wound closure index" a measure of the rate of repair.[97,98] For a given percentage of burn, the chance of survival is greater for younger patients than for elderly, independently of the method used for calculation.

CONCLUSION

In conclusion, the identification of a marker for skin stem cells is a step forward in the field of skin research. A better understanding of the fundamental process involved in skin renewal and regeneration could help design new methods for the treatment of burn patients.

Acknowledgments

The authors are grateful to Drs. François Auger, Francine Goulet (LOEX), Carlos A. López Valle (Hôpital de Chicoutimi), Normand Marceau (Cancer Research Center, Laval University), Pierre Coulombe (Dept. of Biological Chemistry, Johns Hopkins University), Pierrette Gaudreau and André Royal (University of Montreal) for helpful discussions, and the clinicians for their active participation: Drs. Georges

Hébert, Normand Houle, Raymond Labbé, Alphonse Roy, Félix-André Têtu (Hôpital du Saint-Sacrement, Québec) Gabriel Gosselin, Jean-Marc Laliberté (Hôpital Jeffery Hale, Québec) Louise Caouette-Laberge (Hôpital Sainte-Justine, Montréal). The authors want to thank the Medical Research Council of Canada, the "Fonds de la recherche en santé du Québec", the "Fondation des Pompiers du Québec pour les Grands Brûlés" and the "Fondation de l'Hôpital du Saint-Sacrement" for their financial support over the years. Finally, special thanks go to Julie Bergeron, Brigitte Allard, Rina Guignard, Lina Carrier, Thanh-Liem Truong, Nathalie Tremblay for technical assistance and Claude Marin (Hôpital du Saint-Sacrement) for photographic assistance.

References

1. Leblond CP. The life history of cells in renewing systems. Am J Anat 1981; 160:114-157.
2. Oliver RF, Jahoda CAB. The dermal papilla and maintenance of hair growth. In: Rogers GE, Reis PJ, Waid KA, Marshall RC, eds. The Biology of Wool and Hair. New-York: Chapman & Hall, 1988:5,51-67.
3. Pinkus H. Epithelial-mesodermal interaction in normal hair growth, alopecia, and neoplasia. J Dermatol 1978; 5:93-101.
4. Kligman AM. The human hair cycle. J Invest Dermatol 1959; 33:307-316.
5. Chase H B. Growth of the hair. Physiol Rev 1954; 34:113-126.
6. Marceau N, Blouin M-J, Germain L et al. Role of different epithelial cell types in liver ontogenesis, regeneration and neoplasia. In Vitro Cell Dev Biol 1989; 25:336-341.
7. Marceau N, Blouin M-J, Noël M et al. The role of bipotential progenitor cells in liver ontogenesis and neoplasia. In: Sirica AE, ed. The Role of Cell Types in Hepato-Carcinogenesis. Boca Raton: CRC Press Inc., 1992:121-150.
8. Germain L, Blouin M-J, Marceau N. Biliary epithelial and hepatocytic cell lineage relationships in embryonic rat liver as determined by the differential expression of cytokeratins, α-fetoprotein, albumin, and cell surface-exposed components. Cancer Res 1988; 48:4909-4918.
9. Germain L, Noël M, Gourdeau H et al. Promotion of growth and differentiation of rat ductular oval cells in primary culture. Cancer Res 1988; 48:368-378.
10. Sigal SH, Brill S, Fiorino AS Reid L. The liver as a stem cell and lineage system. Am J Physiol 1992; 263:139-148.
11. Brill S, Holst P, Sigal S et al. Hepatic progenitor populations in embryonic, neonatal, and adult liver. Proc Soc Exp Biol Med 1993; 204:261-269.

12. Lendahl U, Zimmerman LB, McKay RDG. CNS stem cells express a new class of intermediate filament protein. Cell 1990; 60:585-595.
13. Price J. Brain stems. The discovery of stem cells in adult nerve tissue makes the repair of brain damage a realistic possibility. Current Biol 1995; 5(3):232-234.
14. Spangrude GJ, Heimfeld S, Weissman IL. Purification and characterization of mouse hematopoietic stem cells. Science 1988; 241:58-62.
15. Thorgeirsson SS. Hepatic stem cells. Am J Pathol 1993; 142:1331-1333.
16. Uchida N, Fleming WH, Alpern EJ et al. Heterogeneity of hematopoietic stem cells. Curr Opin Immunol 1993; 5:177-184.
17. Hall PA, Watt FM. Stem cells: the generation and maintenance of cellular diversity. Development 1989; 106:619-633.
18. Lajtha LG. Stem cell concepts. Differentiation 1979; 14:23-34.
19. Lavker RM, Sun T-T. Heterogeneity in epidermal basal keratinocytes: morphological and functional correlations. Science 1982; 215: 1239-1241.
20. Potten CS, Loeffler M. Stem cells: attributes, cycles, spirals, pitfalls and uncertainties. Lessons for and from the crypt. Development 1990; 110:1001-1020.
21. Argyris TS. Kinetics of epidermal production during epidermal regeneration following abrasion in mice. Am J Pathol 1976; 83:307-316.
22. Eisen AZ, Holyoke JB, Lobitz WC. Responses of the superficial portion of the human pilosebaceous apparatus to controlled injury. J Invest Dermatol 1955; 25:145-156.
23. Stenn DS, DePalma L. Re-epithelialization. In: Clark RAF, Henson PM, eds. The Molecular and Cellular Biology of Wound Repair. New-York: Plenum Press, 1988:321-335.
24. Lenoir MC, Bernard BA, Pautrat G et al. Outer root sheath cells of human hair follicle are able to regenerate a fully differentiated epidermis in vitro. Dev Biol 1988; 130:610-620.
25. Iversen OH, Bierknes R, Devik F. Kinetics of cell renewal, cell migration and cell loss in the hairless mouse dorsal epidermis. Cell Kinet 1968; 1:351.
26. Potten CS, Morris RJ. Epithelial stem cells in vivo. J Cell Sci 1988; 10:45s-62s.
27. Weinstein GD. Epidermal cell kinetics. In: Fitzpatrick TB, Eisen AZ, Wolff K, Freedberg IM, Austen FK, eds. Dermatology in General Medicine. 3rd edition. New York: McGraw-Hill Book Co, 1987: 154-165.
28. Potten CS. Cell replacement in epidermis (keratopoiesis) via discrete units of proliferation. Int Rev Cytol 1981; 69:271-318.
29. Bickenbach JR. Identification and behavior of label retaining cells in oral mucosa and skin. J Dent Res 1981; 60:1611-1620.
30. Morris RJ, Fisher SM, Slaga TJ. Evidence that the centrally and peripherally located cells in the murine epidermal proliferative unit are two distinct cell populations. J Invest Dermatol 1985; 84:277-281.

31. Potten CS. The epidermal proliferative unit: the possible role of the central basal cell. Cell Tissue Kinet 1974; 7:77-88
32. Schmidt GH, Blount MA, Ponder BAJ. Immunochemical demonstration of the clonal organization of chimaeric mouse epidermis. Development 1987; 100:535-541.
33. Madsen A. Studies on the bulge (Wulst) in superficial basal cell epitheliomas. Arch Dermatol 1964; 89:698-708.
34. Mehregan AH, Hashimoto K. Normal structure of the skin. In: Mehregan AH, Hashimoto K, eds. Pinkus' Guide to Dermatohistopathology. San Mateo: Appleton & Lange Press, 1986:5-39.
35. Cotsarelis G, Sun T-T, Lavker RM. Label-retaining cells reside in the bulge area of pilosebaceous unit: implications for follicular stem cells, hair cycle, and skin carcinogenesis. Cell 1990; 61:1329-1337.
36. Lavker RM, Miller S, Wilson C. Hair follicle stem cells: their location, role in hair cycle, and involvement in skin tumor formation. J Invest Dermatol 1993; 101:16s-26s.
37. Morris RJ, Fisher SM, KLein-Szanto AJP et al. Subpopulations of primary adult murine epidermal basal cells sedimented on density gradients. Cell Tissue Kinet 1990; 23:587-602.
38. Pavlovitch JH, Rizk-Rabin M, Jaffray P et al. Characteristics of homogeneously small keratinocytes from newborn rat skin: possible epidermal stem cells. Am J Physiol 1991; 261 (Cell Physiol 30):C964-C972.
39. Rheinwald JG, Green H. Serial cultivation of strains of human epidermal keratinocytes: the formation of keratinizing colonies from single cells. Cell 1975; 6:331-343.
40. Rheinwald JG, Green H. Epidermal growth factor and the multiplication of cultured human epidermal keratinocytes. Nature 1977; 265:421-424.
41. Green H, Barrandon Y. Cultured epidermal cells and their use in the generation of epidermis. NIPS 1988; 3:53-56.
42. Barrandon Y, Green H. Three clonal types of keratinocyte with different capacities for multiplication. Proc Natl Acad Sci USA 1987; 84:2302-2306.
43. Stasiak PC, Purkis PE, Leigh IM. Keratin 19: predicted amino acid sequence and broad tissue distribution suggest it evolved from keratinocyte keratins. J Invest Dermatol 1989; 92:707-716.
44. Lane EB, Wilson CA, Hughes BR et al. Stem cells in hair follicles. Cytoskeletal studies. Ann NY Acad Sci 1991; 642:197-213.
45. Narisawa Y, Hashimoto K, Kohda H. Immunohistochemical demonstration of keratin 19 expression in isolated human hair follicles. J Invest Dermatol 1994; 103:191-195.
46. Coulombe PA, Fuchs E. Elucidating the early stages of keratin filament assembly. J Cell Biol 1990; 111:153-169.

47. Moll R, Franke WW, Schiller DL et al. The catalog of human cytokeratins: Patterns of expression in normal epithelial, tumors and cultured cells. Cell 1982a; 31:11-24.
48. Fuchs E, Green H. Changes in keratin gene expression during terminal differentiation of the keratinocyte. Cell 1980; 19:1033-1042.
49. Osborn M, Weber K. Intermediate filaments: cell-type-specific markers in differentiation and pathology. Cell 1982; 31:303-306.
50. Lane EB, Bartek J, Purkins PE et al. Keratin antigens in differentiating skin. Ann NY Acad Sci 1985; 455:241-258.
51. Sun T-T, Eichner R, Nelson WG et al. Keratin classes: molecular markers for different types of epithelial differentiation. J Invest Dermatol 1983; 81:109s-115s.
52. Fuchs E. Epidermal differentiation: the bare essentials. J Cell Biol 1990; 111:2807-2814.
53. Moll R, Moll I, Wiest W. Changes in the pattern of cytokeratin polypeptides in epidermis and hair follicles during skin development in human fetuses. Differentiation 1982; 23:170-178.
54. Stark H-J, Breitkreutz D, Limat A et al. Keratins of the human hair follicle: "hyperproliferative" keratins consistently expressed in outer root sheath cells in vivo and in vitro. Differentiation 1987; 35:236-248.
55. Heid HW, Moll I, Franke WW. Patterns of expression of trichocytic and epithelial cytokeratins in mammalian tissues. I. Human and bovine hair follicles. Differentiation 1988; 37:137-157.
56. Coulombe PA, Kopan R, Fuchs E. Expression of keratin 14 in the epidermis and hair follicle: insights into complex programs of differentiation. J Cell Biol 1989; 109:2295-2312.
57. Bartek J, Taylor-Papadimitriou J, Miller N et al. Patterns of expression of keratin 19 as detected with monoclonal antibodies in human breast tissues and tumours. Int J Cancer 1985; 36:299-306.
58. Su L, Morgan PR, Thomas JA et al. Expression of keratin 14 and 19 mRNA and protein in normal oral epithelia, hairy leukoplakia, tongue biting and white sponge nevus. J Oral Pathol Med 1993; 22:183-189.
59. Dale BA, Holbrook KA, Kimball JR et al. Expression of epidermal keratins and filaggrin during human fetal skin development. J Cell Biol 1985; 101:1257-1269.
60. Oliver AM. The cytokeratin expression of cultured human foetal keratinocytes. Br J Dermatol 1990; 123:707-716.
61. Michel M, Török N, Godbout M-J et al. Keratin 19 as a biochemical marker of skin stem cells in vivo and in vitro: keratin 19 expressing cells are differentially localized in function of anatomic sites, and their number varies with donor age and culture stage. J Cell Sci 1996; 109:1017-1028.
62. Benoit R, Ling N, Brazeau P et al. Peptides: strategies for antibody production and radioimmunoassays. In: Boulton AA, Baker G.B, Pittman QJ, eds. Neuromethods. Clifton, New Jersey: Humana Press, 1987:43-72.

63. Lussier M, Ouellet T, Lampron C et al. Mouse keratin 19: complete amino acid sequence and gene expression during development. Gene 1989; 85:435-444.
64. Kanedo T, LePage GA. Growth characteristics and drug responses of a murine lung carcinoma in vitro and in vivo. Cancer Res 1978; 38:2084-2090.
65. Bartek J, Bartkova J, Taylor-Papadimitriou J et al. Differential expression of keratin 19 in normal human epithelial tissues revealed by monospecific monoclonal antibodies. Histochem J 1986; 18:565-575.
66. Wilson CL, Dean D, Lane EB et al. Keratinocyte differentiation in psoriatic scalp: morphology and expression of epithelial keratins. Br J Dermatol 1994; 131:191-200.
67. Wolbach SB. The hair cycle of the mouse and its importance in the study of sequences of experimental carcinogenesis. Ann NY Acad Sci 1951; 53:517-536.
68. Lavker RM, Sun T-T. Epidermal stem cells. J Invest Dermatol 1983; 81:121s-127s.
69. Moll R, Moll I, Franke WW. Identification of Merkel cells in human skin by specific cytokeratin antibodies: changes of cell density and distribution in fetal and adult plantar epidermis. Differentiation 1984; 28:136-154.
70. Moll I. Merkel cell distribution in human hair follicles of the fetal and adult scalp. Cell Tissue Res 1994; 277:131-138.
71. Fradette J, Godbout MJ, Michel M et al. Localization of Merkel cells at hairless and hairy human skin sites using keratin 18. Biochem Cell Biol 1995; 73:635-639.
72. Hartschuh W, Weihe E, Yanaihara N. Immunohistochemical analysis of chromogranin A and multiple polypeptides in the mammalian Merkel cell: further evidence for its paraneuronal function? Arch Histol Cytol 1989; 52:423s-431s.
73. Gu J, Polak JM, Tapia FJ et al. Neuron-specific enolase in the Merkel cells of mammalian skin. Am J Pathol 1981; 104:63-68.
74. Rochat, A, Kobayashi K, Barrandon Y. Location of stem cells of human hair follicles by clonal analysis. Cell 1994; 76:1063-1073.
75. Yang JS, Lavker RM, Sun T-T. Upper human hair follicle contains a subpopulation of keratinocytes with superior in vitro proliferative potential. J Invest Dermatol 1993; 101:652-659.
76. Jones PH, Watt FM. Separation of human epidermal stem cells from transit amplifying cells on the basis of differences in integrin function and expression. Cell 1993; 73:713-724.
77. Jones PH, Harper S, Watt FM. Stem cell patterning and fate in human epidermis. Cell 1995; 80:83-93.
78. Kaufmann R, Frösch D, Westphal C et al. Integrin VLA-3: ultrastructural localization at cell-cell contact sites of human cell cultures. J Cell Biol 1989; 109:1807-1815.
79. Lacour JP, Dubois D, Pisani A et al. Anatomical mapping of Merkel cells in normal human epidermis. Br J Dermatol 1991; 125:535-542.

80. Germain L, Rouabhia M, Guignard R et al. Improvement of human keratinocyte isolation and culture using thermolysin. Burns 1993; 2:99-104.
81. Barrandon Y, Green H. Cell migration is essential for sustained growth of keratinocyte colonies: the roles of transforming growth factor-α and epidermal growth factor. Cell 1987a; 50:1131-1137.
82. Gilchrest BA. In vitro assessment of keratinocyte aging. J Invest Dermatol 1983; 81:181s-189s.
83. Germain L, Auger FA. Tissue engineered biomaterials: biological and mechanical characteristics. In: Wise D, Trantolo DJ, Altobelli DE, Yaszemski MJ, Gresser JD, Schwartz ER, eds. Encyclopedic Handbook of Biomaterials and Bioengineering, Part B; Applications. New York: Marcel Dekker, Inc., 1995:699-734.
84. Bell E, Erhlich HP, Buttle DJ et al. A living tissue formed in vitro and accepted as a full thickness skin equivalent. Science 1981; 211:1042-1054.
85. Saintigny G, Bonnard M, Damour O et al. Reconstruction of epidermis on a chitosan cross-linked collagen-GAG lattice: effect of fibroblasts. Acta Derm Venereol 1993; 73:175-180.
86. Harriger MD, Warden GD, Greenhalgh DG et al. Pigmentation and microanatomy of skin regenerated from composite grafts of cultured cells and biopolymers applied to full-thickness burn wounds. Transplantation 1995; 59(5): 702-707.
87. Fleischmajer R, Contard P, Schwartz E et al. Elastin-associated microfibrils (10 nm) in a three dimensional fibroblast culture. J Invest Dermatol 1991; 97:638-643.
88. Michel M, Germain L, Bélanger PM et al. Functional evaluation of anchored skin equivalent cultured in vitro: Percutaneous absorption studies and lipid analysis. Pharm Res 1995; 12(3):455-458.
89. Auger FA, Lopez Valle CA, Guignard R et al. Skin equivalent produced with human collagen. In Vitro Cell Develop Biol Animal 1995; 531:432-439.
90. Berthod F, Germain L, Guignard R et al. In vitro reexpression of collagen XIV and small splice variant of collagen XII in a human reconstructed skin and their in vivo differential expression in human skin. J Invest Dermatol 1996. In revision.
91. Naughton GK, Jacob L, Naughton B. A physiological skin model for in vitro toxicity studies. In: Goldberg A, ed. In Vitro Toxicity: New Directions. New York: Liebert, 1989:183-189.
92. López Valle CA, Germain L, Rouabhia M et al. Grafting on nude mice of living skin equivalents produced using human collagens. Transplantation 1996; 62(3):317-323.
93. Michel M, L'Heureux N, Auger FA et al. From newborn to adult: phenotypic and functional properties of skin equivalent and human skin as a function of donor age. J Cell Phys 1996; in press.

94. Grove GL. Age-related differences in healing of superficial skin wounds in humans. Arch Dermatol Res 1982; 272:381-385.
95. Levenson SM, Lund CC. Dermatome skin grafts for burns in patients prepared with dry dressings and with and without penicillin. New Engl J Med 1945; 233:607-611.
96. Levenson SM, Birkhill FR, Waterman DF. The healing of soft tissue wounds: the effect of nutrition, anemia, and age. Surgery 1950; 28:905-935.
97. Scott-Conner CEH, Meydrech E, Wheeler WE et al. Quantitation of rate of wound closure and the prediction of death following major burns. Burns 1988; 14(5):373-378.
98. Scott-Conner CEH, Coil Jr JA, Conner HF et al. Wound closure index: a guide to prognosis in burned patients. J Trauma 1986; 26(2):123-127.
99. Stern M, Waisbren BA. Comparison of methods of predicting burn mortality. Burns 6:119-125, 1978.
100. Laemmli UK. Cleavage of structural proteins during the assembly of the head of bacteriophage T4. Nature 1970; 227:680-685.
101. O'Farrell PZ, Goodman HM, O'Farrell PH. High resolution two-dimensional electrophoresis of basic as well as acidic proteins. Cell 1977; 12:1133-1142.
102. Towbin H, Stacheli T, Gordon J. Electrophoretic transfer of proteins from polyacrylamide gels to nitrocellulose sheets: procedures and some applications. Proc Natl Acad Sci USA 1979; 76:4350-4354.
103. Marceau N, Germain L, Goyette R et al. Cell of origin of distinct cultured rat liver epithelial cells, as typed by cytokeratin and surface component selective expression. Biochem Cell Biol 1986; 64:788-802.

CHAPTER 9

In Vitro Applications of Skin Substitutes for Dermatological Purposes

François Berthod and François A. Auger

INTRODUCTION

What was deemed to be impossible barely two decades ago is now a reality. The new and exciting methods ushered in by the tissue engineering approaches now allow the reconstruction of human skin in vitro albeit as a simplified tissue: no adnexa are present in such substitutes. However, these skin models have already proven to be extraordinary tools for various in vitro applications. The present chapter will give an overview of such utilizations since 1993 in keeping with the scope of this book.

IN VITRO MODELS OF TISSUE ENGINEERED SKIN SUBSTITUTES

Tissue engineering, as explained in the previous chapters, allows the regeneration of a tissue or an organ in the laboratory. The first stage of this process is the expansion in a very efficient manner of the initial cells obtained from a biopsy. Thereafter the recombination of these cellular elements within the appropriate environment and according to precise culture conditions will provide a nearly inexhaustible source of living tissue substitute.

The in vitro development of a human skin substitute has been the focus of several studies for at least 40 years. The major purpose of

Skin Substitute Production by Tissue Engineering: Clinical and Fundamental Applications, edited by Mahmoud Rouabhia.

these experiments was the production of a large amount of autologous skin for the coverage of deep and extensive wounds (see chapter 2).[1] A first step has been accomplished in the engineering of skin with the production of cultured epithelial sheets since the early 1980s (see chapters 2 and 3). A second step has now been achieved in the laboratory: the engineering of both the dermal and epidermal skin layers. The structural foundation of all skin substitutes is dependent on the dermal layer upon which the epidermal cells are to be seeded. Therefore, the reconstruction of the dermis in vitro necessitates the use of fibroblasts since they are the central cellular element of the dermis. Striking differences have been shown between the behavior of fibroblasts cultured on two-dimensional substrates and cells cultured in a three-dimensional environment. When cultured in monolayer, fibroblasts display high proliferation rate and high collagen synthesis which are not consistent with the in vivo situation, while their mitotic behavior and their collagen production are inhibited when they are embedded in a collagen gel.[2,3] Thus, the culture of fibroblasts in a three-dimensional system is required to re-establish the in vivo environment of these cells. In addition, it allows the reproduction of the structure and thickness of the human dermis. The scaffold of the dermal equivalent in which fibroblasts are cultured is a key element that largely governs specific properties and features of the different skin models. It can be made of a collagen solution in which cells are embedded or it can be composed of a hard collagenous or synthetic substrate into which cells spread and migrate.[1] The following paragraphs will illustrate these three main approaches to tissue engineering of skin substitutes. In all instances the description will insist on the dermal component. Indeed, the reconstruction of the dermis in vitro is the pivotal step in skin tissue production since the epidermal layer is always obtained in the same fashion: seeding keratinocytes on its surface.

In all of these skin substitutes a more complete epidermal differentiation can be obtained by raising these constructs at an air-liquid interface. This is done after a variable time period of submerged culture to allow full coverage of the dermal component by the keratinocytes previously seeded. The contact with air of the upper layer of epidermis, in addition to creating a nutrient pathway through the dermal layer, will induce many aspects of the keratinization process found in skin (Figs. 9.1 and 9.2).

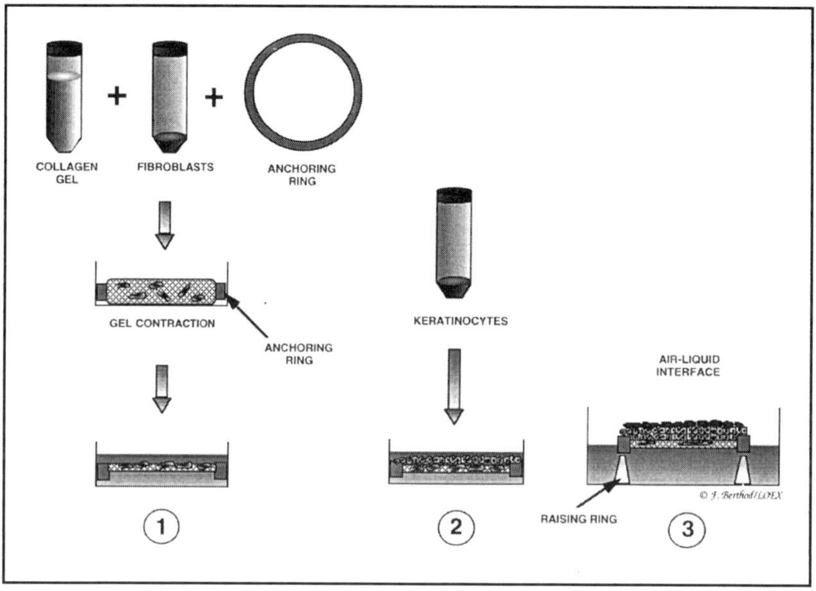

Fig. 9.1. Preparation of an anchored collagen gel skin substitute. Human foreskin fibroblasts are mixed with a collagen gel and poured in a 35 mm Petri dish in which an anchoring ring of paper is placed. The anchoring ring prevents gel contraction in diameter. Following a maturation period of the gel, keratinocytes are seeded on the gel and the substitute is raised to the air-liquid interface to promote epidermal differentiation (see also Fig. 9.2).

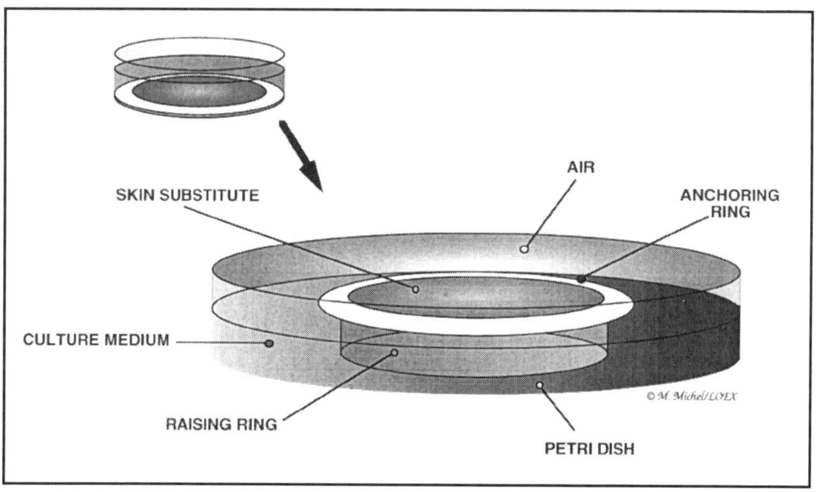

Fig. 9.2. Raising of skin substitutes at the air-liquid interface. The anchored collagen gel can be raised at the air-liquid interface by putting it on a plastic ring which support the paper ring of the anchored collagen gel.

Skin Substitute Made From Collagen Gel

Since the major protein found in human dermis is collagen and since collagen is easily extracted from rat tail tendons, it has been readily used to produce the first model of dermal equivalent. This model has been developed by Bell in 1979 following previous works.[4] Briefly, fibroblasts are embedded in a collagen gel at the appropriate physicochemical conditions which then recreates the dermal layer. This approach takes advantage of the spontaneous ability of collagen monomers to rearrange into fibrils at neutral pH, and of the contractile properties of fibroblasts upon these collagen fibrils.[4] The ensuing tissue contraction is proportional to the cell number and inversely proportional to the collagen gel concentration. Finally, keratinocytes are seeded on top of these gels in order to create the epidermal layer. Until a few years ago the company Organogenesis Inc. (Canton, MA) was offering such a skin model for in vitro studies.

The collagen gel system has been extensively studied and is undoubtedly the most popular dermal equivalent used in the literature probably because it is easy to produce. In addition, fibroblasts cultured in a gel closely mimic the in vivo situation: they display similar shape, mitotical activity and collagen synthesis compared to human dermis. Paradoxically, the main advantage and the major drawback of this model is the contraction of the gel by fibroblasts. This phenomenon usually leads to a loss of about 80% of the initial surface.[2,5] Indeed, if the retraction process is useful in the study of wound contraction, it becomes a major inconvenience when a clinical application is desired or when a high reproducibility of samples is required. We devised a solution to this problem: the collagen gel can be poured in a 5 mm-wide paper filter ring to which the gel attaches.[5] These anchored gels allow the preservation of the exact initial surface with only a decrease in thickness (Fig. 9.1). Furthermore, histological analysis showed an alignment of fibroblasts and collagen fibers resembling normal dermal tissue. Lastly this anchorage facilitates manipulations for other in vitro applications.[6]

Skin Substitute Made From Collagen Sponge

The collagen sponge, which was previously developed for clinical application in burn wounds (see chapter 4), offers interesting features for in vitro applications in dermatology.[7] As the name implies its dermal component is composed of a lyophilized collagen mixture which presents a porous structure into which the seeded fibroblasts can creep. Thereafter, keratinocytes are deposited on the upper planar surface formed by these fibroblasts. The collagen sponge system is more readily amenable to industrial production than is the collagen gel construct. This process is briefly described in Figure 9.3.[8] In contrast with the collagen gel model, the collagen network of the sponge prevents any contraction of the dermal equivalent in diameter or thickness, thus allowing the production of more reproducible samples. In addition, fibroblasts cultured in a collagen sponge display a completely different behavior compared with fibroblasts embedded in a collagen gel, as for example in terms of protein synthesis. In the first weeks of culture, fibroblasts exhibit moderate mitotic activity and high protein synthesis approaching the level found in the healing process, while after one month their protein production and proliferation profile is markedly reduced mimicking the in vivo situation.[2,3] The major advantage of this model is the production of a high amount of newly synthesized extracellular matrix that fills the sponge's pores after one month (Fig. 9.4).[3,9] This is a very interesting feature because this extracellular matrix is made of various types of molecules secreted by the fibroblasts in response to their environment rather than the "instantly" provided matrix found in the collagen gel system. For example, we have demonstrated that fibroblasts cultured in the collagen sponge re-expressed type XIV collagen and the small splice variant of type XII collagen, although they lose their

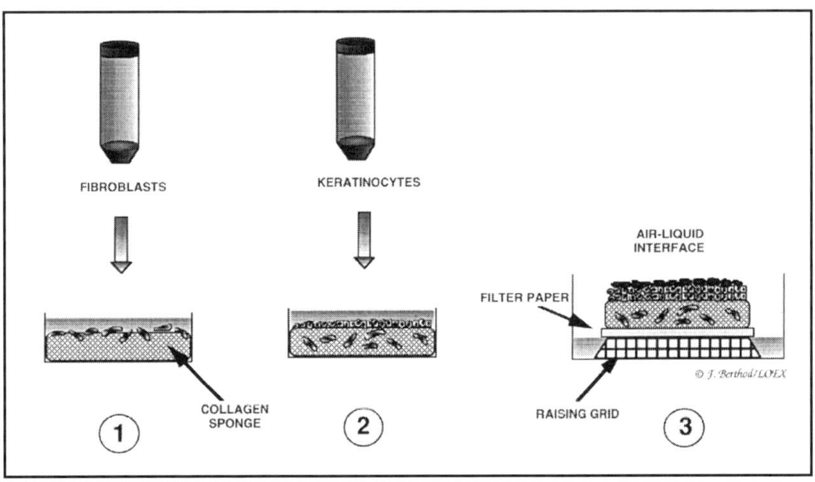

Fig. 9.3. Preparation of the collagen sponge. After mixing of type I and III bovine collagen, chitosan and chondroitin 4-6 sulfate, the solution is poured in 6 wells plates, frozen at -70°C and then lyophilized.

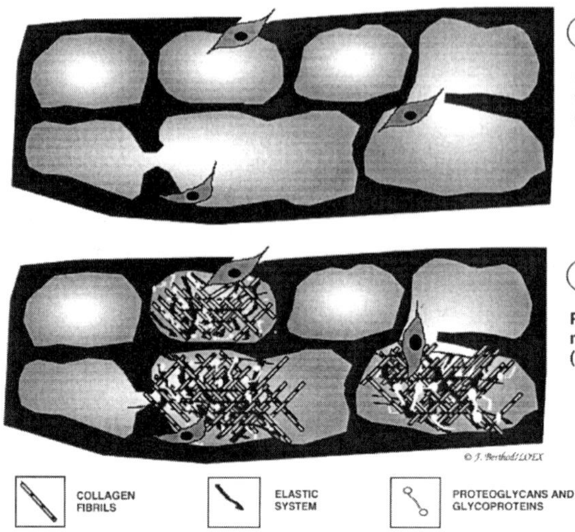

Fig. 9.4. Schematic representation showing the migration of fibroblasts through the porous network of a collagen sponge skin substitute, followed after 1 month by the filling of pores with extracellular matrix, and stabilization of protein production.

ability to produce these proteins when cultured in monolayer on plastic and in collagen gel.[10] Furthermore, the fibrillogenesis process that leads to the formation of collagen fibrils necessitates the presence of mature collagen in order to progress, while fibroblasts cultured in vitro are known to mostly synthesize procollagen. As we have previously demonstrated,[3] the high amount of collagen molecules produced in the collagen sponge spontaneously polymerizes to form quarter-staggered collagen fibrils, thus showing that mature collagen is synthesized in this model. Furthermore, these fibrils are associated in interwoven bundles that mimic the fibrillar organization of the dermis.[3] The remodeling of a highly differentiated extracellular matrix in the collagen sponge is a key feature in the reconstruction of these skin substitutes (Fig. 9.5). The processes of collagen maturation, fibrillogenesis, and fibrils organization in bundles are of outstanding importance to reconstruct an extracellular matrix that reproduces the normal dermis. In addition, a well-differentiated extracellular matrix greatly favors the process of epidermal differentiation, as shown in Figure 9.6, and also the formation of a nicely organized dermal-epidermal junction.[11,12]

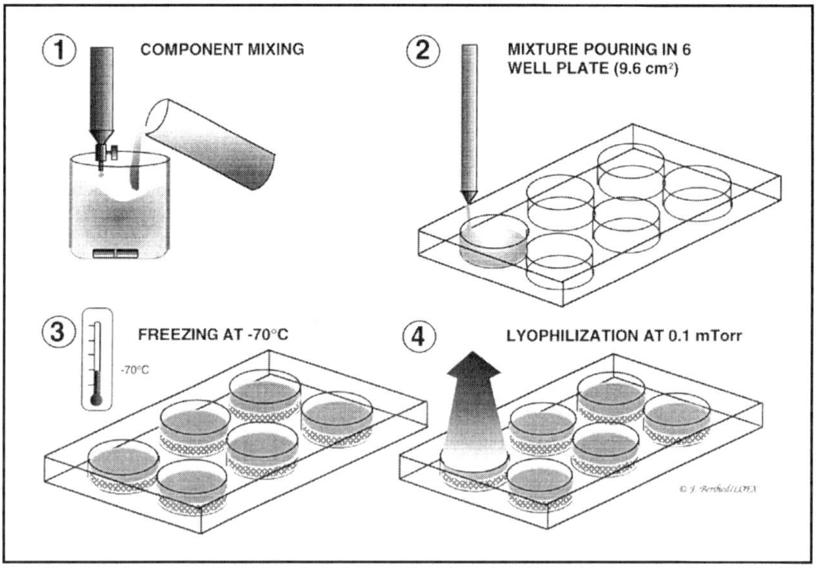

Fig. 9.5. Preparation of the collagen sponge skin substitute. After rehydration of the sponge in culture medium, fibroblasts are seeded on its top, followed by keratinocytes a few days later. The substitute is put on a paper filter and then lifted to the air-liquid interface on a stainless-steel grid, in order to promote epidermal differentiation.

Fig. 9.6. Histological cross-section of a collagen sponge skin substitute stained by Masson's trichrome. A well organized epidermis with a thick stratum corneum is obtained after 20 days of maturation at the air-liquid interface. Note the sponge network in black, in which fibroblasts deposit a dense extracellular matrix, which serves as a substrate for keratinocytes adhesion.

Skin Substitute Made From Synthetic Mesh

Human three-dimensional skin substitutes (Skin^2 ZK1000, ZK1100, ZK1200, ZK1300) have been developed by the Advanced Tissue Sciences Corporation (La Jolla, CA) by culturing human neonatal foreskin fibroblasts onto a Nylon mesh, and afterward covering them with keratinocytes (models ZK1200 and ZK1300 only). In this skin model, fibroblasts grow and produce their own collagen, growth factors and extracellular matrix proteins inside this mesh, while keratinocytes form a multilayered epidermis. Skin^{2}ZK1200 is made of human skin epithelial-fibroblast cocultures which display a noncornified epithelium similar to cornea and conjunctiva. This substitute is used for the evaluation of ocular irritancy.[13] Skin^{2}ZK1300 promotes the formation of a fully differentiated epidermis and is devoted to skin corrosion testing.[14] Model ZK1300 is a 11x11 mm skin substitute cultured into a MILLICELL™ culture insert (Millipore, Bedford, MA) in 6-well plates. Cells are fed with serum-free DMEM medium added beneath the culture insert.[14] One of the major advantages of these models was their commercial availability throughout the world (Fig. 9.7). Unfortunately, this company has stopped production of all their in vitro skin substitutes in 1996.

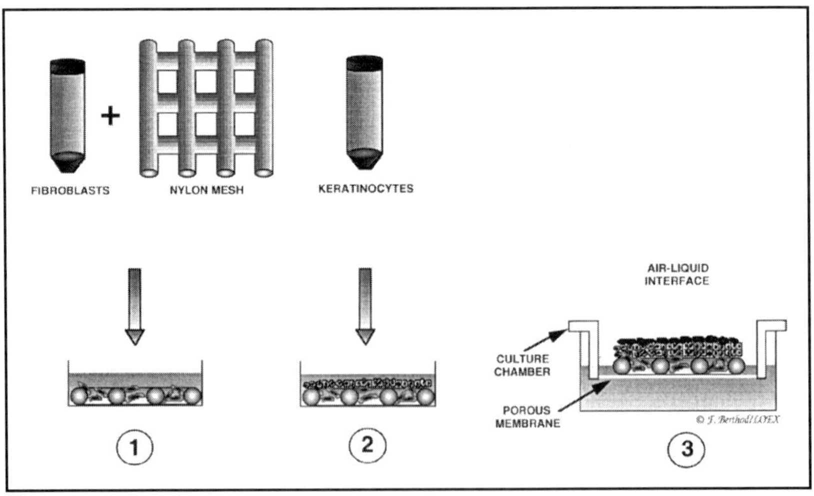

Fig. 9.7. Preparation of the Nylon mesh skin substitute. Fibroblasts are seeded on a Nylon mesh, and after a maturation period covered with keratinocytes. Then they are raised to the air-liquid interface by transferring the mesh in a MILLICELL™ chamber, to promote epidermal differentiation.

APPLICATION OF SKIN SUBSTITUTES IN CUTANEOUS PHYSIOLOGY

The skin is a complex organ that contains several different types of cells and molecules in addition to an intricate three-dimensional architecture. Skin appendages are deeply anchored in the dermis where their vascular and nervous plexuses are intermingled. Thus the previously described skin substitutes are tools of outstanding interest for the study of cutaneous physiology. But the reconstructed skin is a much simpler tissue in which only the main cellular and molecular components are present, i.e., keratinocytes, fibroblasts and collagen. Furthermore, it must be noted that the results obtained are not modified by other cell types or humoral modulation. This allows for untainted observations on one hand but a less physiological environment on the other hand.

Skin Substitutes in the Study of Cell/Extracellular Matrix Interaction

Since fibroblasts cultured on a two-dimensional substrate display high mitotic activity and protein synthesis, it seems likely that their purpose is to fill their environment as fast as possible with new extracellular matrix. This behavior is typical of the wound healing process in which cells must quickly replenish tissue loss after an injury. This process is governed in vivo by inflammation, and in vitro, by fetal calf serum that contains a high amount of growth factors and hormones. The process is arrested by suppressor factors from the immune system in vivo, while in vitro, the activating signal originating from serum will be maintained throughout the culture period. Nevertheless, the mitotic activity and protein synthesis displayed in vitro by fibroblasts also decrease with time, as we previously demonstrated.[3] In vitro the suppressor signal originates from the extracellular matrix which induces a feedback inhibition on collagen synthesis, as it closely surrounds cells, and also an inhibition of their mitotical activity. However, the mechanism whereby contact with extracellular matrix molecules triggers intracellular modulation of gene expression remains unclear. To date the most plausible hypothesis could be the fixation of extracellular matrix proteins, such as collagens, to integrins which induces post-transcriptional modification of intracellular proteins via tyrosine phosphorylation. Indeed, change in phosphorylation correlated with increased activity of specific protein kinase has been observed following three-dimensional contact with type I collagen.[15] In addition, it is noteworthy that the feedback control induced by the surrounding extracellular matrix on cellular metabolism is mediated by specific molecules. For example, fibroblasts cultured in a fibrin gel display high collagen synthesis and mitotic activity although they are surrounded by fibrin fibrils, in contrast with the behavior of fibroblasts cultured in a collagen gel.[16]

Skin Substitutes in the Study of Cell/Cell Interaction

The role of dermis in the cutaneous physiology has been neglected for a long time. This skin layer was considered to be a simple provider of nutritive elements for the epidermis while imparting elastic properties to the skin. Indeed, it was very difficult to assess the eventual modulation of epidermal differentiation by fibroblasts in vivo. The use of tissue engineered skin substitutes now renders such cell/cell interactions much easier to investigate.

Is it really necessary to add living fibroblasts in a skin substitute, knowing that it will be a source of additional cost and trouble? This question has been raised by several investigators. Tuan et al have shown that dermal fibroblasts activate keratinocyte outgrowth on collagen gels,[17] while Saintigny et al[11] demonstrated that fibroblasts promote the differentiation of a multilayered stratified epithelium, covered with a compact stratum corneum, in contrast with disorganized epithelial cell layers in the absence of fibroblasts. In addition, mesenchymal-epithelial interactions are essential in the dermal-epidermal junction formation. Indeed, fibroblasts are the only source of nidogen, the linkage protein of the basal lamina, during dermal-epidermal junction formation.[18] Fibroblasts also contribute to the synthesis of type IV and VII collagens and laminin[19] which are the main component of the dermal-epidermal junction, while enhancing keratinocyte metabolic activity as we demonstrated.[20] Even more fascinating is the ability of fibroblasts to induce an increase of type VII and kalinin gene expression by keratinocytes.[12] When a direct contact is allowed between fibroblasts and keratinocytes, a continuous lamina densa with anchoring fibrils is observed at very early stages of culture, while none of these structures are seen in absence of direct cell contact with fibroblasts, although all the components previously cited are present.[11,12] In contrast, keratinocytes can also modulate the biosynthetic phenotype of fibroblasts at a pretranslational level.[21] Thus, fibroblasts induce an increased synthesis of dermal-epidermal junction constituents, and most of all, they promote the correct organization of these components into the characteristic form of a lamina densa.

Skin Substitutes in the Study of Dermal Tensile Properties

Skin substitutes have facilitated an analysis of the precise role of each of the different components, cellular and extracellular, in the overall tensile properties of human skin. The tensile properties of skin substitutes have to be evaluated as a function of their structural components in order to give them the best mechanical resistance and stretchability characteristics. Indeed, the excessive fragility of a skin substitute which does not allow its easy handling is a frequent and major drawback. As we previously demonstrated, the evaluation of the mechanical property of a biomaterial can be a major parameter in its optimization.[8] Indeed, in the preparation of our collagen sponge biomaterial, the ratio chitosan/collagen was critical to strengthen the tensile property of the sponge as monitored using an extensometer.[8] Different methods have been recently developed to evaluate the tensile properties depending upon the type of skin substitutes analyzed. A model collagen gel system has been used to study the effect of macromolecular components of the extracellular matrix on the viscoelastic properties of dermis in the absence of cells. Thus, dermatan sulfate proteoglycan, hyaluronic acid as well as the amount of elastin and the collagen fibril diameter can modify the viscoelastic behavior of the collagen gel model.[22] More surprisingly, Nishiyama et al[23] also demonstrated that collagen molecules themselves, i.e., type XII and XIV collagens, may modulate extracellular matrix deformability by increasing hydrated collagen fibrils mobility within an acellular collagen gel. However, decorin, a small dermatan sulfate proteoglycan produced by fibroblasts, has been shown to inhibit collagen gel contraction by glycosaminoglycan-glycosaminoglycan interactions.[24] Of course, such molecules are produced in vivo by fibroblasts, which remain the bandmaster of the extracellular matrix modeling and contraction, except when it is mostly driven by inflammation. Since fibroblasts themselves may usually strengthen the mechanical property of the skin substitute in which they are cultured, specific techniques should be developed to evaluate the mechanical property of living skin substitutes. Thus, the physical response of fibroblast-populated collagen gels to tensile strain has been evaluated with a servo-controlled linear actuator.[25] In addition, we have developed an indentation apparatus to study the tensile properties of living skin substitutes according to their composition (Fig. 9.8).[26]

APPLICATION OF SKIN SUBSTITUTES IN CUTANEOUS PATHOLOGY

Skin Substitutes in the Study of Dermatological Diseases

Skin diseases can originate from either the dermis or the epidermis. The causal role of each layer in the physiopathology of cutaneous diseases can be explored by various permutations of diseased and normal cells in both the epidermal and/or the dermal component of tissue engineered skin substitutes. Such models are unique tools to carefully analyze disease-induced alterations in cell function and extracellular matrix, and thus, help find drugs to correct these anomalies.

A good example of the use of skin substitutes in the study of dermatological diseases is psoriasis. The morphological and biochemical characteristics of psoriasis are well known, but the pathogenesis of the disease is not well documented. Skin substitutes models of psoriasis do not completely reproduce a stable psoriatic phenotype since

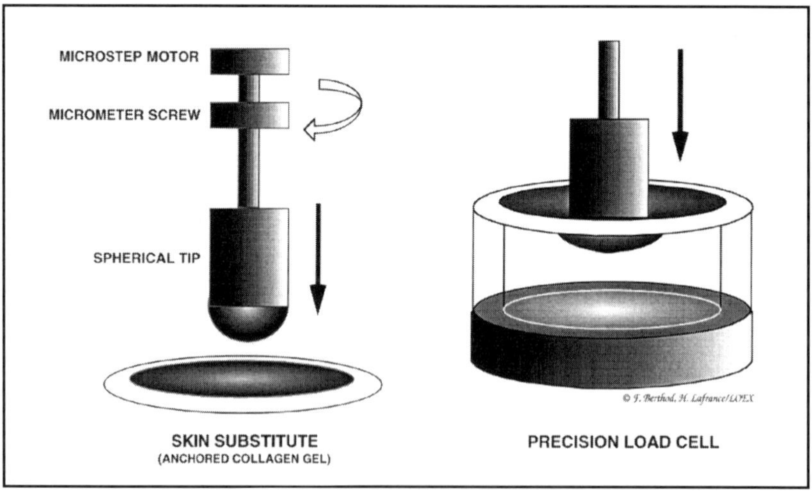

Fig. 9.8. Schematic representation of the indentation apparatus for tensile properties evaluation of skin substitutes. Vertical displacement of spherical tip induces axisymmetric tension within the anchored collagen gel skin substitute installed on the tissue support holder and resulting tension within skin equivalent is measured by a precision load cell.[26]

they need to be maintained by extracutaneous stimuli.[27] However they may still be a good model which resembles psoriatic skin.[28,29] Accordingly, Konstantinova et al[30] demonstrated that psoriatic fibroblasts produce increased amount of interleukin-8 in a skin substitute. Interleukin-8 may play a major role in the determination of psoriatic phenotype, by a paracrine loop between fibroblasts and keratinocytes.

Skin substitutes are also well adapted to the study of ichtyotic epidermal cells, since such substitutes usually show an intrinsic tendency for keratinocyte hyperproliferation. Major features of ichtyotic keratinocytes, such as hyperkeratosis and hyperproliferation, appear to be retained when these cells are cultured in a skin substitute.[31,32]

Skin Substitutes in the Study of Wound Healing

Wound healing models using skin substitutes

The in vivo wound healing phenomenon is so complex that simpler in vitro models should shed some light into some of its controversial aspects.

An interesting in vitro wound healing model has been developed by Stephens et al.[33] They prepared in vitro a normal fibroblast-populated collagen gel, combined or not with an epidermal coverage, and then created a punch biopsy wound in this dermal equivalent. The wounded gel was then transferred onto an acellular collagen lattice in order to facilitate its contraction. The repopulation of the "wound space" could be easily monitored, in parallel with its re-epithelialization. Normal fibroblasts have also been embedded in a fibrin gel to mimic wound repair.[16] This model allowed matrix contraction, cell growth and collagen production by fibroblasts, in contrast with a floating collagen gel in which cell proliferation and collagen synthesis were inhibited. The specific traction force exerted by fibroblasts on collagen gel has been studied by different models, such as the fibroblast-populated collagen microsphere assay[34,35] and the culture force monitor for measurement of contraction forces.[36,37] In another set of experiments, Genever et al[38] compared normal fibroblasts embedded in floating collagen gel to fibroblasts grown in attached collagen matrices. Fibroblasts cultured in attached collagen gels were more sensitive to phenytoin (used as an inducer of connective tissue growth) and seemed to more closely mimic a wound healing situation.

Another approach in the development of a wound healing model is the use of fibroblasts directly extracted from wounds. Thus, Calderon et al have developed a wound model in which fibroblasts are obtained from keloids, and are cultured in monolayer on plastic. Wounding is then created by a linear cut.[39] Finally, our group produced an in vitro wound healing model consisting in wound fibroblasts embedded in a floating collagen gel.[40] This model permitted the monitoring of gel contraction, as well as the characterization of its intracellular components by immunocytochemistry.[40,41] Since wound fibroblasts display greater contractile properties than normal fibroblasts this model should mimic more closely the wound healing process (Fig. 9.9).

Skin substitutes to study the modulation of wound healing by inflammation factors

Growth factors are the major components involved in enhancement and control of wound healing. Some of them increase the wound contraction as well as the extracellular matrix production while others do the opposite. Such regulations were clearly demonstrated using skin substitutes. TGF-β1 enhances contraction of normal fibroblasts-

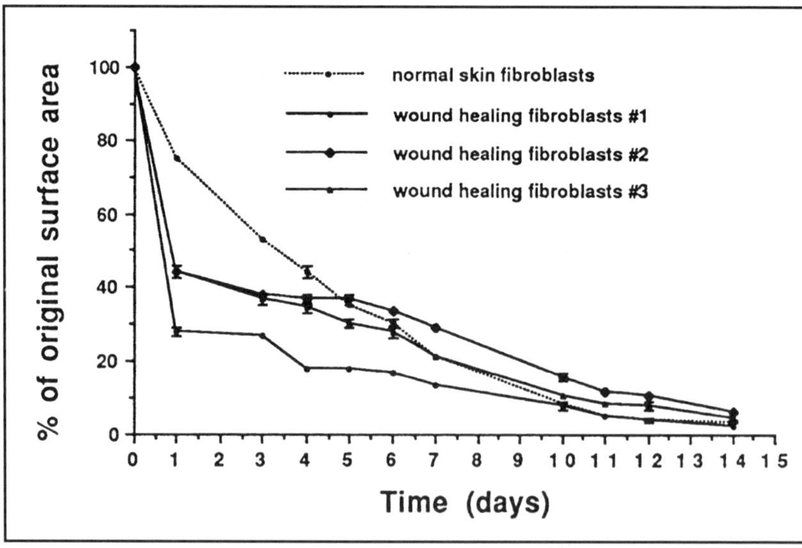

Fig. 9.9. Floating collagen gel contraction by normal and wound healing human dermal fibroblasts. Note that the extent of contraction is greater when wound healing fibroblasts are used compared with dermal fibroblasts. Reprinted with permission from Germain L et al, J Surg Res 1994; 57:268-273.

populated collagen gels and type I collagen synthesis[42,43] together with fibroblast-populated fibrin matrix[16] while addition of TGF-β neutralizing antibodies to the culture medium inhibits fibroblast-induced collagen gel contraction.[44] In contrast, interferon-α, -β, -γ have been shown to inhibit fibroblast-populated collagen gel contraction, and thus should be good candidates to reduce the scarring phenomenon associated with severely contracted wound in vivo.[45]

Skin substitutes to study the process of wound contraction

The contractile property of the dermis is a key element in the esthetic and functional result of wound healing following an injury. If wound contraction is useful to accelerate healing by drawing together wound edges, it may lead to excessive scar formation and pathological wound contracture. For example, the retraction process may greatly impair joint functionality through excessive contracture.

Skin substitutes, and especially the collagen gel dermal equivalent, are very well suited to study similar dermal contractile phenomenons. The collagen gel dermal equivalent is based upon the process of cell/matrix contraction. Indeed, when fibroblasts are embedded in a type I/III collagen gel, they spontaneously induce a contraction of collagen fibrils until the physiological density of the extracellular matrix is reached, i.e., 70 mg/ml.[46] This process has been extensively studied and gives collagen gel equivalents their strength.

What is the exact process explaining wound contraction?

There are no doubt that this fibroblastic process is modulated at least in part by inflammation. As discussed above, inflammation signaling is transmitted to cells by growth factors and cytokines.

What is the fibroblast's response to this signaling?

Two main theories have been proposed, one privileging an increase in fibroblast locomotion, the other emphasizing a change in cell phenotype, from fibroblast to myofibroblast.[40] Since fibroblasts extracted from wounds display greater contractile properties compared with normal fibroblasts (Fig. 9.9),[41] the latter hypothesis seems to describe more closely the wound healing process. Indeed, our group demonstrated that 20-80% of wound fibroblasts contain smooth muscle α-actin in contrast to only 10% of normal fibroblasts.[40] Smooth muscle α-actin is a more specific feature of myofibroblasts and is a

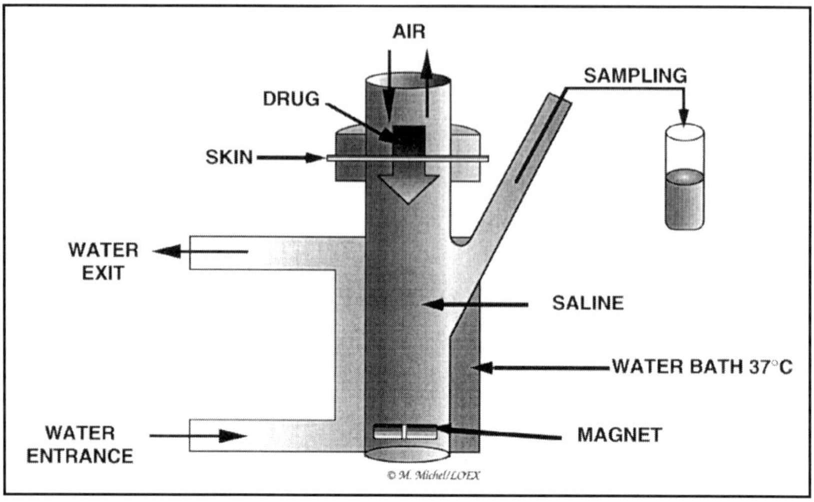

Fig. 9.10. Schematic representation of the Franz diffusion cell used for evaluation of drug permeability. Radioactive drug is put on the top of the skin substitute and its diffusion is monitored by measuring the radioactivity release for 24 h.[49,51]

key element in collagen gel contraction.[47] Thus, it is noteworthy that the wound fibroblast population may be constituted of a great proportion of myofibroblasts although a clear myofibroblast phenotype is still difficult to describe. Originating from fibroblast, their phenotype may be switched to myofibroblast by induction of inflammation factors such as TGF-β although this hypothesis is still controversial.

The mechanism by which contractile forces are generated from fibroblasts has also been extensively studied. Therefore, it has been shown recently that the fibroblasts induced collagen gel contraction is mediated by $\alpha_2\beta_1$ and, to a lesser extent, by $\alpha_1\beta_1$ integrin complexes.[48] These integrins allow the binding of collagen fibrils to intracellular cytoskeleton via a transmembrane receptor. This contraction phenomenon is also mediated by several growth factors, such as TGF-β, EGF, or PDGF.[43]

APPLICATION OF SKIN SUBSTITUTES IN CUTANEOUS PHARMACOLOGY

Skin Substitutes in the Study of Drug Permeability and Biotransformation Through Skin

Transdermal delivery of drugs has recently generated a lot of interest. In vitro testing of drug diffusion through skin is one of the first step in transdermal drug development. Different models have been designed to evaluate drug permeability in vitro instead of experimentation on living animals. Since these models must use fresh or frozen animal and human skin, this situation may lead to technical and ethical problems.[49] Tissue engineered skin substitute are imperfect models for the exact quantitative permeability evaluation of compounds since their barrier function is much lower than human or animal skin.[50] Nevertheless their qualitative value is very promising since the percutaneous absorption profiles of various drugs were comparable to those obtained with human and animal skin. Thus, even though skin substitutes are very permeable, they still allow a valuable comparative analysis of various compounds.[49,51] Furthermore, the optimization of the culture conditions has led to a significant improvement of the barrier function of these skin substitute in a very promising way.[52] Indeed, using the Franz cell diffusion technique (Fig. 9.10), our group demonstrated a relationship between lipid type and content, and the percutaneous absorption rate in a skin substitute made of a anchored fibroblast-populated collagen gel covered by a well differentiated epidermis (Fig. 9.11).[49,51] Thus, the lower ceramide density in the substitute epidermis likely explained its higher permeability.

In the process of drug development, the phenomenon of their biotransformation during diffusion through skin must also be investigated.[53,54] This analysis demands living and metabolically active skin cells as the ones encountered in tissue engineered skin substitutes.

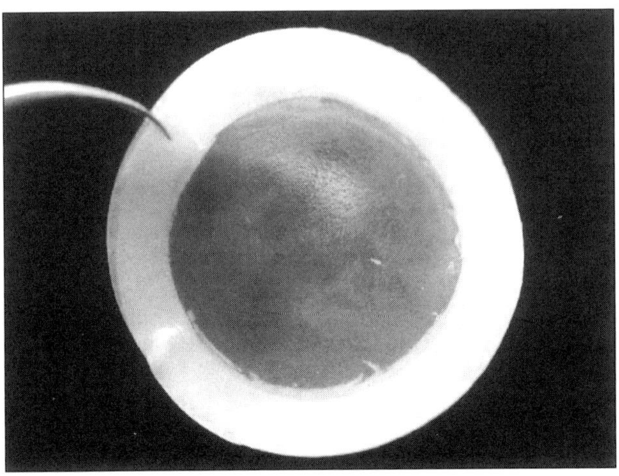

Fig. 9.11. Macroscopic view of an anchored collagen gel skin substitute used for drug permeability assessment.

For example, cyclosporin A has been shown to be less efficient as a topical treatment compared to an oral administration in the treatment of psoriasis. The use of the human skin model made of fibroblasts cultured in a Nylon mesh demonstrated that a high biotransformation of cyclosporine A in the skin induced a lower efficacy of the compound in a topical application.[54]

In conclusion, skin substitutes are unique tools in the study of drug permeability and biotransformation. Although skin substitutes do not reproduce until now a barrier function as efficiently as human skin, they allow the study of drug bioavailability and biotransformation on living human fibroblasts and keratinocytes, in contrast with nonliving frozen human skin, or fresh animal skin. In addition, skin substitutes, when prepared on a large scale basis, should be more reproducible than human or animal skin which present high interindividual disparities.

Skin Substitutes in the Study of Drug Effect on Skin

Modern pharmacological and dermatological research requires appropriate skin substitutes to allow a simple, rapid and low cost screening of new drugs developed by chemists. Many of the studies of drug effect on skin are performed on well established "normal skin,"[55] wound healing and psoriasis models. Indeed, extensive work regarding the effect of retinoic acid has been performed on skin substitutes.[56-58] Retinoic acid is used in the treatment of psoriasis to decrease hyperkeratosis, but also seems to promote protective effects against UVB-induced damages to skin (see chapter 7). Lastly, retinoic acid could decrease the formation of contractile scars by inhibiting wound contraction, as shown on fibroblasts-populated collagen gels.[58]

Skin substitutes allow a better comprehension of the subtle and various mechanisms by which retinoic acid exerts such pleiotropic effects on skin. By modulating gene transcription after binding to nuclear receptors, this molecule was shown in skin model to provoke metaplasia of epithelium,[56] suppression of loricrin expression,[57] as well as changes in actin cytoskeleton of fibroblasts.[58] Drug effect on wound healing has also been extensively studied using skin substitutes. As previously described, inflammation factors can modulate wound healing and scar formation, while drugs such as phenytoin inhibits collagenase activity and stimulates cell proliferation in an anchored fibroblast-populated collagen gel.[59]

APPLICATION OF SKIN SUBSTITUTES IN CUTANEOUS TOXICOLOGY

Background Information

Skin is the first protective barrier of our organism against a multitude of potentially toxic agents. This includes therapeutics agents, personal hygiene preparations, cosmetics, household cleaners, industrial solvents, pesticides, fertilizers, paints, machine lubricants, to name a few. Skin damage potential, such as corrosion, of these chemicals has been the focus of an extensive regulatory effort. Thus the Code of Federal Regulations[60] or the OECD[61] testing guidelines state the proper classification, labeling, packaging, and transportation according to their respective toxicity. Traditionally, skin toxicity assessments involve topical application of chemicals to the skin of rabbits. The production of irreversible full-thickness necrosis of the skin is determined by visual inspection of the skin for up to 21 days following exposure. The major drawbacks of skin corrosion testing on animals are:

1. a subjectivity of scoring,
2. an unpredictable correlation between animals and human of the results obtained,
3. a high cost and long time requirements of testing methods,
4. the severe pain inflicted to animals which becomes more and more ethically debatable.

These drawbacks are spurring the interest of industry and government into alternative in vitro methods. As a first step, tissue engineers developed new in vitro models to assess skin corrosion, and as a second step, governments validated these models and procedures in new official testing guidelines.

Use of Skin Substitutes

Different skin substitutes have been developed to assess toxicity in vitro. Although toxicity testing has been evaluated in vitro on mesenchymal and epithelial monolayer cultures[62] or on dermal equivalents,[63-65] it is of critical importance to develop skin substitutes displaying a well differentiated epidermis attached to a fibroblast-populated dermal equivalent. These models will allow a more sophisticated approach to in vitro skin toxicity evaluation. Furthermore, in an attempt to standardize toxicity assays worldwide following international regulations, it seems preferable to produce industrially a skin substitute as a "kit format" or to develop a standardized model easy to reproduce in any tissue culture laboratory. Collagen gel based skin substitute has been used to produce a eye irritation model using fibroblast and keratinocyte cell lines. This model seems to be easy to produce since it is made with common cell lines and it is based on the well known collagen gel technology.[66] The Skin^{2}Model ZK1200 developed by Advanced Tissue Sciences is undoubtedly the most popular skin substitute for the assessment of skin toxicity through the world.[14,67-69] Furthermore, the Skin^{2}Model ZK1350 skin corrosion test have been granted US Department of Transportation approval.[69] An interlaboratory evaluation of this model, compared with animal skin (TER test) and a synthetic biobarrier membrane (CORROSITEX™) have been performed by the European Centre for Validation of Alternative Methods toward various corrosives substances. The results obtained using the Skin^2 test showed no significant interlaboratory differences, while the concordance between the in vitro and in vivo chemical classifications ranged from 70-92%. Thus, this test is promising in terms of its ability to identify corrosives and noncorrosives.[69] But as previously stated these models are no more in production. Another model made in the form of an industrial kit is currently being developed using the fibroblast-populated collagen sponge as dermal equivalent.[63,64] This model allow a high reproducibility in the cutaneous toxicity response and should be commercialized soon.

Toxicity Assessment In Vitro

The procedures used to assess cutaneous toxicity on skin substitutes consist in cell morphology, cell growth and cell viability analysis. Cell growth can be evaluated by measurement of the cell number or the DNA content (by ^{3}H-thymidine uptake) of skin substitutes. The most popular procedure to measure cell viability use MTT,[13,14,63-65,68,70] a biochemical marker of mitochondrial metabolism, while neutral red incorporation by intracellular lysosomes is also used alone or in conjunction with MTT.[13,14] MTT staining of cellular viability is a simple, rapid and low cost procedure in which the purple coloration of cell lysate is proportional to cell viability, and is measured by spectrophotometry (Figs. 9.12, 9.13). Other types of cytotoxicity evaluation can be performed, such as measurement of intracellular lactate deshydrogenase release following membrane damage, or ATP

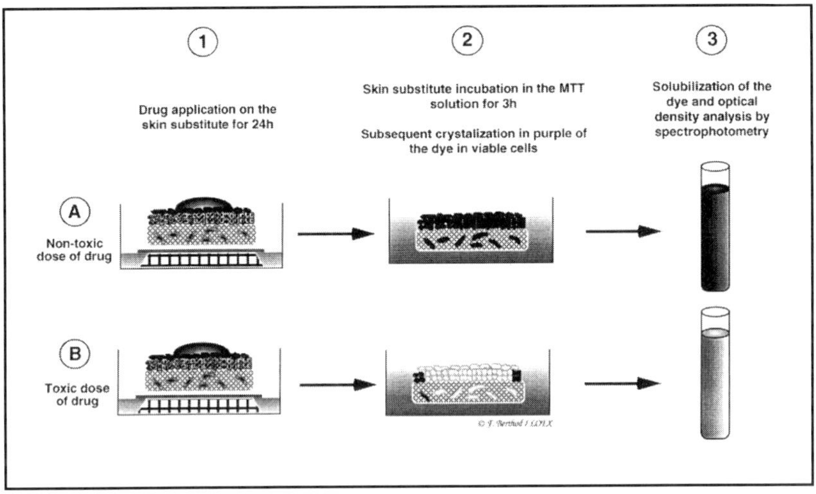

Fig. 9.12. Schematic representation of the MTT staining procedure to evaluate drug toxicity using skin substitutes. The drug is applied on the epidermal layer of the substitute for 24 h. Samples are then incubated in the MTT solution for 3 h. Viable cells are subsequently stained in purple by precipitation of MTT in formazan crystals, while dead cells are not stained. Formazan crystals are then solubilized in isopropyl alcohol. The optical density of samples is analyzed by spectrophotometry and the percentage of viability is calculated by reference to a 100% viable control.

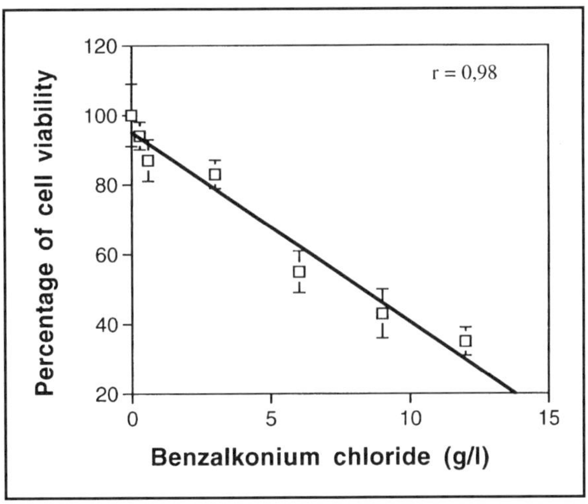

Fig. 9.13. Benzalkonium chloride effects on cell viability. The percentage of cell viability in the collagen sponge was evaluated following skin substitutes incubation with increasing concentrations of benzalkonium chloride deposited at the top of the epidermal layer. The toxicity is calculated using the MTT staining procedure. Note the linearity of the substitute response to the toxic agent and the low dispersion of the values. Bars indicate standard deviation of four samples.

content analysis,[66] as well as cytokines and prostaglandin E_2 release to monitor inflammatory mediator cell production following irritation.[14,63]

Future Developments

Finally, skin substitutes are expected to give accurate predictions of the skin corrosivity potential of chemicals. Work remains to be done to explore further the "border-line" areas between corrosive and noncorrosive components. Skin substitutes have not allowed until now the study of more complex reactions to irritants, such as inflammatory or allergic reactions, due to the absence of immunocompetent cells in the models. Nevertheless, immunologically active skin substitutes are currently under development and should be of great interest for cutaneous toxicity assessment.[71] Thus, skin substitutes need to be greatly improved before all animal testing can be eliminated. But they are valuable tools that could yet serve as a prescreen, prior to carrying out animal tests for the classification of new chemicals.

CONCLUSION

Tissue engineered skin substitutes have opened many new approaches in the assessment of cutaneous physiology, pathology, pharmacology and toxicology. Because of the very nature of this method of tissue reconstruction, many unanswered questions can now be addressed. Furthermore these procedures are amenable to a high level of reproducibility which is of paramount importance if they are to be widely accepted. In all event a lot of the future development of such in vitro methodology is very dependent upon regulation concerning the use of animals in skin research and drug development. Such an impetus was clearly the source of very significant efforts in Europe after the announcement, such regulation prohibiting animal use in cosmetological research. We think that the field of in vitro application of skin substitutes for dermatological research has a bright future because of this combination of intrinsic value and extrinsic demand.

Acknowledgments

The authors are grateful to Drs. Lucie Germain, Francine Goulet (LOEX), Carlos A. López Valle (Hôpital de Chicoutimi), for helpful discussions, and the clinicians for their active participation: Drs.

Georges Hébert, Normand Houle, Alphonse Roy, Félix-André Têtu (Hôpital du Saint-Sacrement, Québec). The authors want to thank the Medical Research Council of Canada, the "Fonds de la recherche en santé du Québec", the "Fondation des Pompiers du Québec pour les Grands Brûlés" and the "Fondation de l'Hôpital du Saint-Sacrement" for their financial support over the years. Finally, special thanks go to Julie Bergeron, Rina Guignard, Lina Carrier, Nathalie Tremblay, Kathleen Baker for technical assistance and Claude Marin (Hôpital du Saint-Sacrement) for photographic assistance.

References

1. Berthod F, Damour O. In vitro reconstructed skin models for deep burn wound coverage. Brit J Dermatol. In press.
2. Berthod F, Hayek D, Damour O, Collombel C. Collagen synthesis by fibroblasts–cultured within a collagen sponge. Biomaterials 1993; 14:749-754.
3. Berthod F, Sahuc F, Hayek D. Deposition of collagen fibril bundles by long-term culture of fibroblasts in a collagen sponge. J Biomed Mater Res 1996; 32:87-94.
4. Bell E, Erhlich HP, Buttle DJ et al. A living tissue formed in vitro and accepted as a full thickness skin equivalent. Science 1981; 211:1042-1054.
5. Auger FA, Lopez Valle CA, Guignard R et al. Skin equivalent produced with human collagen. In Vitro Cell Dev Biol Anim 1995; 31:432-439.
6. Germain L, Auger FA. Tissue engineered biomaterials: biological and mechanical characteristics. In: Wise DL, Trantolo DJ, Altobelli DE et al, eds. Encyclopedic Handbook of Biomaterials and Bioengineering, Part B: Applications. New York: Marcel Dekker, Inc. Publishers, 1995; 1:699-734.
7. Yannas IV, Lee E, Orgill DP et al. Synthesis and characterization of a model extracellular matrix that induces partial regeneration of adult mammalian skin. Proc Natl Acad Sci USA 1989; 86:933-937.
8. Berthod F, Saintigny G, Chretien F et al. Optimization of thickness, pore size and mechanical properties of a biomaterial designed for deep burn coverage. Clin Mater 1994; 15:259-265.
9. Middelkoop E, de Vries HJ, Ruuls L et al. Adherence, proliferation and collagen turnover by human fibroblasts seeded into different types of collagen sponges. Cell Tissue Res 1995; 280:447-453.
10. Berthod F, Germain L, Guignard R et al. Collagen XIV and small and large splice variants of collagen XII are differently expressed in vivo in human skin, and are expressed in vitro in a human reconstructed skin. J Invest Dermatol; in press.

11. Saintigny G, Bonnard M, Damour O, Collombel C. Reconstruction of epidermis on a chitosan cross-linked collagen-GAG lattice: effect of fibroblasts. Acta Dermatol Venereol 1993; 73:175-180.
12. Sahuc F, Nakazawa K, Berthod F et al. Mesenchymal-epithelial interactions regulate gene expression of type VII collagen and kalinin in keratinocytes and dermal-epidermal junction formation in a skin equivalent model. Wound Rep Reg 1996; 4:93-102.
13. Osborne R, Perkins MA, Roberts DA. Development and intralaboratory evaluation of an in vitro human cell-based test to aid ocular irritancy assessments. Fund Appl Toxicol 1995; 28:139-153.
14. Whalen E, Donnely TA, Naughton G, Rheins LA. The development of three-dimensional in vitro human tissue models. Hum Exp Toxicol 1994; 13:853-859.
15. Roeckel D, Krieg T. Three-dimensional contact with type I collagen mediates tyrosine phosphorylation in primary human fibroblasts. Exp Cell Res 1994; 211: 42-48.
16. Tuan TL, Song A, Chang S et al. In vitro fibroplasia: matrix contraction, cell growth, and collagen production of fibroblasts cultured in fibrin gels. Exp Cell Res 1996; 223:127-134.
17. Tuan TL, Keller LC, Sun D et al. Dermal fibroblasts activate keratinocyte outgrowth on collagen gels. J Cell Sci 1994; 107: 2285-2289.
18. Fleischmajer R, Schechter A, Bruns M et al. Skin fibroblasts are the only source of nidogen during early basal lamina formation in vitro. J Invest Dermatol 1995; 105:597-601.
19. Marinkovich MP, Keene DR, Rimberg CS, Burgeson RE. Cellular origin of the dermal-epidermal basement membrane. Dev Dynam 1993; 197:255-267.
20. Goulet F, Poitras A, Rouabhia M et al. Stimulation of human keratinocyte proliferation through growth factor exchanges with dermal fibroblasts in vitro. Burns 1996; 22:107-112.
21. Lacroix M, Bovy T, Nusgens BV, Lapiere CM. Keratinocytes modulate the biosynthetic phenotype of dermal fibroblats at a pretranslational level in a human skin equivalent. Arch Dermatol Res 1995; 287:659-664.
22. Hsu S, Jamieson AM, Blackwell J. Viscoelastic studies of extracellular matrix interactions in a model native collagen gel system. Biorheology 1994; 31:21-36.
23. Nishiyama T, McDonough AM, Bruns RR, Burgeson RE. Type XII and XIV collagens mediate interactions between banded collagen fibers in vitro and may modulate extracellular matrix deformability. J Biol Chem 1994; 45:28193-28199.
24. Bittner K, Liszio C, Blumberg P et al. Modulation of collagen gel contraction by decorin. Biochem J 1996; 314:159-166.
25. Özerdem B, Tözeren A. Physical response of collagen gels to tensile strain. J Biomech Eng 1995; 117:397401.

26. Lafrance H, Guillot M, Germain L, Auger FA. A method for the evaluation of tensile properties of skin equivalents. Med Eng Phys 1995; 17:537-543.
27. Mils V, Basset-Seguin N, Moles JP et al. Comparative analysis of normal and psoriatic skin both in vivo and in vitro. Differentiation 1994; 58:77-86.
28. Leigh IM, Navsaria H, Purkis PE et al. Keratins (K16 and K17) as markers of keratinocyte hyperproliferation in psoriasis in vivo and in vitro. Br J Dermatol 1995; 133:501-511.
29. van Ruissen F, de Jongh GJ, Zeeuwen PL et al. Induction of normal and psoriatic phenotypes in submerged keratinocyte cultures. J Cell Phys 1996; 168:442-452.
30. Konstantinova NV, Duong DMT, Remenyik E et al. Interleukin-8 is induced in skin equivalents and is highest in those derived from psoriatic fibroblasts. J Invest Dermatol 1996; 107:615-621.
31. Featherstone J, Dykes PJ, Marks R. The growth of ichthyotic epidermal cells in a 3-dimensional reconstruction of human skin, the skin equivalent. Clin Exp Dermatol 1995; 20:313-318.
32. Amsellem C, Haftek M, Hoyo E et al. Evidence of increased keratinocyte proliferation in air-liquid interface culture of non-bullous congenital ichthyosiform erythroderma. Acta Dermatol Venereol 1993; 73:262-269.
33. Stephen P, Wood EJ, Raxworthy MJ. Development of a multilayered in vitro model for studying events associated with wound healing. Wound Rep Reg 1996; 4:393-401.
34. Moon AG, Tranquillo RT. Fibroblast-populated collagen microsphere assay of cell traction force: Part 1. Continuum model. AIChE J 1993; 39:163-177.
35. Barocas VH, Moon AG, Tranquillo RT. The fibroblast-populated collagen microsphere assay of cell traction force. Part 2: Measurement of the cell traction parameter. J Biomech Eng 1995; 117:161-170.
36. Eastwood M, McGrouther DA, Brown RA. A culture force monitor for measurement of contraction forces generated in human dermal fibroblast cultures: evidence for cell-matrix mechanical signaling. Biochim Biophys Acta 1994; 1201:186-192.
37. Eastwood M, Porter R, Khan U et al. Quantitative analysis of collagen gel contractile forces generated by dermal fibroblasts and the relationship to cell morphology. J Cell Physiol 1996; 166:33-42.
38. Genever PG, Wood EJ, Cunliffe WJ. The wounded dermal equivalent offers a simplified model for studying wound repair in vitro. Exp Dermatol 1993; 2:266-273.
39. Calderon M, Lawrence WT, Banes AJ. Increased proliferation in keloid fibroblasts wounded in vitro. J Surg Res 1996; 61:343-347.
40. Moulin V, Castilloux G, Jean A et al. In vitro models to study wound healing fibroblasts. Burns 1996; 22:359-362.

41. Germain L, Jean A, Auger FA, Garrel DR. Human wound healing fibroblasts have greater contractile properties than dermal fibroblasts. J Surg Res 1994; 57:267-273.
42. Reed MJ, Vernon RB, Abrass IB, Sage EH. TGF-beta 1 induces the expression of type I collagen and SPARC, and enhances contraction of collagen gels, by fibroblasts from young and aged donors. J Cell Physiol 1994; 158:169-179.
43. Piscatelli SJ, Michaels BM, Gregory P et al. Fetal fibroblast contraction of collagen matrices in vitro: the effects of epidermal growth factor and transforming growth factor-beta. Ann Plast Surg 1994; 33:38-45.
44. Pena RA, Jerdan JA, Glaser BM. Effects of TGF-beta and TGF-beta neutralizing antibodies on fibroblast-induced collagen gel contraction: implications for proliferative vitreoretinopathy. Invest Ophth Vis Sci 1994; 35:2804-2808.
45. Dans MJ, Isseroff R. Inhibition of collagen lattice contraction by pentoxifylline and interferon-alpha, -beta, and -gamma. J Invest Dermatol 1994; 102:118-121.
46. Grinnell F, Lamke R. Reorganization of hydrated collagen lattices by human skin fibroblasts. J Cell Sci 1984; 66:51-53
47. Arora PD, McCulloch CA. Dependence of collagen remodeling on alpha-smooth muscle actin expression by fibroblasts. J Cell Physiol 1994; 159:161-175.
48. Carver W, Molano I, Reaves TA et al. Role of the alpha-1 beta-1 integrin complex in the collagen gel. J Cell Physiol 1995; 165:425-437.
49. Michel M, Auger FA, Germain L. Anchored skin equivalent cultured in vitro: a new tool for percutaneous absorption studies. In Vitro Cell Dev Biol Anim 1993; 29:834-837.
50. Roy SD, Fujiki J, Fleitman JS. Permeabilities of alkyl p-aminobenzoates through living skin equivalent and cadaver skin. J Pharm Sci 1993; 82:1266-1268.
51. Michel M, Germain L, Belanger PM, Auger FA. Functional evaluation of anchored skin equivalent cultured in vitro: percutaneous absorption studies and lipid analysis. Pharm Res 1995; 12:455-458.
52. Fartasch M, Ponec M. Improved barrier structure formation in air-exposed human keratinocyte culture systems. J Invest Dermatol 1994; 102:366-374.
53. Kubota K, Ademola J, Maibach HI. Simultaneous diffusion and metabolism of betamethasone 17-valerate. J Pharm Sci 1995; 84:1478-1481.
54. Vickers AE, Biggi WA, Dannecker R, Fischer V. Uptake and metabolism of cyclosporin A and SDZ IMM 125 in the human in vitro skin2 dermal and barrier function models. Life Sci 1995; 57:215-224.
55. Rashid A, Edward M, Richardson MD. Activity of terbinafine on Trichophyton mentagrophytes in a human living skin equivalent model. J Med Vet Mycol 1995; 33:229-233.

56. Asselineau D, Darmon M. Retinoic acid provokes metaplasia of epithelium formed in vitro by adult human epidermal keratinocytes. Differentiation 1995; 58:297-306.
57. Brown LJ, Geesin JC, Rothnagel JA et al. Retinoic acid suppression of loricrin expression in reconstituted human skin cultured at the liquid-air interface. J Invest Dermatol 1994; 102:886-890.
58. Ho-Sung L, Tsuchihashi M, Tenshin S, Katawata T. Effect of retinoic acid on contraction of collagen gel induced by fibroblasts. Biochem Biophys Res Com 1994; 205:455-459.
59. Genever PG, Cunliffe WJ, Wood EJ. Influence of the extracellular matrix of fibroblast responsiveness to phenytoin using in vitro wound healing models. Br J Dermatol 1995; 133:231-235.
60. Anon. Code of Federal Regulations. Method of testing corrosion to the skin. Transportation title. Washington, DC: Office of the Federal Register, National Archives and Records Administration. 1991; 49: Part 173.136, Appendix A.
61. Anon. Acute dermal irritation/corrosion. OECD's Guidelines for the testing of chemicals (Adopted by the Council of July 17, 1992) Paris. OECD, 1992, Specific Guideline Number 404.
62. Dover R, Otto WR, Nanchahal J, Riches DJ. Toxicity testing of wound dressing materials in vitro. Br J Plast Surg 1995; 48:230-235.
63. Augustin C, Damour O. Pharmacotoxicological applications of an equivalent dermis: three measurements of cytotoxicity. Cell Biol Toxicol 1995; 11:167-171.
64. Augustin C, Damour O. Development of kit for predicting cutaneous toxicity in vitro using three-dimensional dermal equivalent: Phase 1, reproductibility of dermal equivalent. Cell Eng 1995; 1:58-62.
65. Lindsay CD, Upshall DG. The generation of a human dermal equivalent to assess the potential contribution of human dermal fibroblasts to the sulphur mustard-induced vesication response. Hum Exp Toxicol 1995; 14:580-586.
66. Pasternak AS, Miller WM. First-order toxicity assays for eye irritation using cell lines: parameters that affect in vitro evaluation. Fund Appl Toxicol 1995; 25:253-263.
67. Pierard GE, Goffin V, Hermanns-LE T et al. Surfactant-induced dermatitis: comparison of corneosurfametry with predictive testing on human and reconstucted skin. J Am Acad Dermatol 1995; 33:462-469.
68. Perkins MA, Osborne R, Johnson GR. Development of an in vitro method of skin corrosion testing. Fund Appl Toxicol 1996; 31:9-18.
69. Botham PA, Chamberlain M, Barratt MD et al. A prevalidation study on in vitro skin corrosivity testing. ATLA 1995; 23:219-255.
70. Osborne R, Perkins MA. An approach for development of alternative test methods based on mechanisms of skin irritation. Food Chem Toxicol 1994; 32:133-142.
71. Rouabhia M. In vitro production and transplantation of immunologically active skin equivalents. Lab Invest 1996; 75:503-517.

CHAPTER 10

SUMMARY AND PERSPECTIVES

Mahmoud Rouabhia

Four to five decades ago, the appropriate dressing for partial and severe skin loss was a major challenge for all medical corps. Indeed, the vast majority of these patients died from their direct injury or injuries related to fire problems. Fortunately, over the last ten years, rescuing and keeping alive patients with over 80% of total surface body damaged skin has become a reality. This progress is due to the appropriate clinical intervention and successful biological dressing. The most valuable example of successful biological dressing is the in vitro engineered autologous epidermis. Indeed, by 1982, two burned children were rescued and kept alive until now by using engineered autologous epidermis. This biological dressing method is considered the most effective life-saving therapy for patients with massive burns. However, just as other therapies, the engineered autologous epidermal substitutes present several disadvantages such as the absence of other crucial epidermal cells (LC, melanocytes) and the lack of the dermis which is an essential part of normal skin, a more effective cutaneous tissue with its protective and also immunological function. Consequently, a large number of scientists around the world are working on engineering a skin substitute that is structurally and functionally close to natural skin.

Natural skin is multicellular and possesses a well-defined three-dimensional architecture. As part of tissue engineering, there is an emerging recognition that these characteristics, as well as others, are critically important to the generation of specific tissues with biological function. For skin, and certainly for some other tissues, structural function and mechanical properties are important. A key requirement may be the presence of a specific extracellular matrix (ECM). In addition, there will need to be both chemical and structural stability, not just in terms of the handling required for implantation, but more importantly for cells to remain viable, once grafted, within the in vivo environment in which the construct must operate. Thus, the core functional element of a tissue engineered skin substitute is live, functional cells.

The major properties that engineered skin should have are:

1. Viable cells
2. The presence of ECM-like or appropriately secreted ECM
3. Cell-cell cooperation
4. Cell-ECM interaction
5. Biosynthesis and secretion of proteins in response to specific stimuli
6. Significant cell functionality for a sufficient period of time
7. Cell availability to engineer these biological tissues
8. Therapeutic application of these engineered tissues for burn and other cutaneous disaster treatment
9. Engineered tissues should be biocompatible and less immunogenic when they are allo- or xenotransplanted.

As tissue engineering is the most recent biological field, some appropriate conditions are already optimized. Indeed, the use of dissociated living cells appropriate or semi-appropriate extracellular matrix, and well-defined in vitro cultured conditions are presently well controlled parameters. These advances made possible the development of biological epidermal substitutes. The long-term goal of engineering these tissues is the development of implantable live tissue, the permanent replacement of damaged skin and the adequate remodeling of definitive generated cutaneous tissues. Although the engineered epidermis is functionally active tissue and clinically useful, some serious disadvantages still hamper the therapeutic use of these biological materials and scientists still have a lot to do.

It is a well-known reality that the tissue engineering field is two decades behind the recombinant DNA industry; however, tissue engineering is evolving rapidly, and it is reasonable to believe that at the end of the present century this multidisciplinary field will reach the size of today's recombinant DNA industry and perhaps beyond.

Indeed, by applying chimeric culture technology to burn treatment, medical corps will be able to reduce the waiting time of these fragile patients for their essential barrier recovery. They will also reduce hospitalization costs. Furthermore, the incorporation of other cutaneous elements (dermis, ECM proteins, immune cells, etc.) to engineered tissues will certainly help to generate more structured, aesthetic and physiologically functional skin. Our perspectives are to continue working on the chimeric culture technology to answer several questions that could be asked such as: What is the optimum proportion of syngeneic keratinocytes needed to generate in vivo functional skin after grafting engineered chimeric epidermis? Is there immune activation of the chimeric implant recipient, and at what level? What will happen structurally and immunologically after two, three or more chimeric substitute grafting to the same recipient? What about a tricellular (syngeneic, allogeneic and xenogeneic) chimeric implant grafting? The answers to these questions and many more may be developed in an updated volume probably with some clinical results using chimeric engineered and grafted human skin substitutes.

Finally, we are aware that the most successful short-term and long-term clinical results could be achieved if engineered skin contains both dermal and epidermal structures. Valuable efforts are presently made in our laboratory, and certainly in other laboratories around the world, to engineer skin that more closely resembles natural skin. In conclusion, tissue engineering is an emerging field. This multidisciplinary and interdisciplinary field offers enormous challenges to engineers and bioscientists. Its development will certainly stem from the productive interactions between multidisciplinary teams involving engineers, life scientists and clinicians.

INDEX

S

T

U

V

W

X